Engineering Electromagnetism

WILEY STUDENT SERIES IN ELECTRONIC AND ELECTRICAL ENGINEERING

Professor C. M. Snowden
*Department of Electronic
and Electrical Engineering
University of Leeds
UK*

Dr A. McCowen
*Department of Electrical
and Electronic Engineering
University College of Swansea
UK*

Design and Technology of Integrated Circuits
D. de Cogan

An Introduction to Applied Electromagnetism
C. Christopoulos

Engineering Electromagnetism
A. J. Baden Fuller

Engineering Electromagnetism

A. J. Baden Fuller
University of Leicester UK

JOHN WILEY & SONS
Chichester · New York · Brisbane · Toronto · Singapore

Copyright © 1993 by John Wiley & Sons Ltd,
Baffins Lane, Chichester,
West Sussex PO19 1UD, England

All rights reserved.

No part of this book may be reproduced by any means, or transmitted, or translated into a machine language without the written permission of the publisher.

Other Wiley Editorial Offices

John Wiley & Sons, Inc., 605 Third Avenue,
New York, NY 10158-0012, USA

Jacaranda Wiley Ltd, G.P.O. Box 859, Brisbane,
Queensland 4001, Australia

John Wiley & Sons (Canada) Ltd, 22 Worcester Road,
Rexdale, Ontario M9W 1L1, Canada

John Wiley & Sons (SEA) Pte Ltd, 37 Jalan Pemimpin #05-04,
Block B, Union Industrial Building, Singapore 2057

Library of Congress Cataloging-in-Publication Data

Baden Fuller, A. J.
   Engineering electromagnetism / A. J. Baden Fuller.
     p.   cm. — (Wiley student series in electronic and electrical engineering)
   Includes bibliographical references and index.
   ISBN 0 471 93489 5
   1. Telecommunication systems.   2. Electromagnetic theory.
   3. Electromagnetic waves—Transmission.   I. Title.   II. Series.
TK5102.5.B294   1993
621.382—dc20   92-44451
                                                                                          CIP

British Library Cataloguing in Publication Data

A catalogue record for this book is available from the British Library

ISBN 0 471 93489 5

Typeset in 10/12pt Palatino by Keytec, Bridport, Dorset
Printed and bound in Great Britain by
Dotesios Ltd, Trowbridge, Wiltshire.

Contents

Preface vii
Acknowledgements ix

Chapter 1 **Electromagnetic Fields** 1
    1.1   Introduction 1
    1.2   Field quantities 2
    1.3   Electromagnetic field relationships 5
    1.4   Vector analysis 10
    1.5   Maxwell's equations 23
    1.6   Sources and potentials 29

Chapter 2 **Transmission Lines I: Switching Surges** 34
    2.1   Introduction 34
    2.2   Transmission line equations 36
    2.3   Line termination 41
    2.4   Voltage step 45

Chapter 3 **Transmission Lines II: A.C. Effects** 58
    3.1   A.C. steady state 58
    3.2   Standing waves 69
    3.3   Terminating conditions 77
    3.4   Impedance measurement 86
    3.5   Impedance matching 91

Chapter 4 **Plane Waves** 101
    4.1   Wave propagation 101
    4.2   Field components of the plane wave 106
    4.3   Reflection and refraction from a plane boundary: (I) normal incidence 112
    4.4   Reflection and refraction from a plane boundary: (II) oblique incidence 116
    4.5   Attenuation 124
    4.6   Conducting medium 128

Chapter 5 **Waveguide Effects** 135
    5.1   Parallel plate waveguide 135
    5.2   Wave velocities 142
    5.3   Transmission line fields 146
    5.4   Waveguide 152

Chapter 6 Hollow Metal Waveguide ... 156
- 6.1 Rectangular waveguide ... 156
- 6.2 Field components in rectangular waveguide ... 163
- 6.3 Dominant mode in rectangular waveguide ... 167
- 6.4 Circular waveguide ... 172
- 6.5 Dominant mode in circular waveguide ... 180
- 6.6 Resonant cavity ... 186

Chapter 7 Optical Fibre ... 193
- 7.1 Surface wave ... 193
- 7.2 Dielectric film waveguide ... 199
- 7.3 Circular fibre ... 206
- 7.4 Attenuation and dispersion ... 214
- 7.5 Directional coupler ... 219

Chapter 8 Radiation and Antennas ... 224
- 8.1 Short dipole antenna ... 224
- 8.2 Antenna arrays ... 233
- 8.3 Long antennas ... 244
- 8.4 Aperture antennas ... 252
- 8.5 Reflector antennas ... 260

Chapter 9 Systems ... 264
- 9.1 Line communications ... 264
- 9.2 Radio-wave propagation ... 265
- 9.3 Microwave communication ... 268
- 9.4 Satellite communication ... 273
- 9.5 Radar ... 277

Bibliography ... 283

Appendix 1 Physical constants ... 285

Appendix 2 Notation ... 286

Index ... 289

Preface

Electrical science pervades the whole of modern civilisation. Its study may be divided into two interlinked disciplines: circuit theory and electromagnetic field theory. Courses in elementary electromagnetic theory study the phenomenon of electrostatics and magnetism and the low-frequency interactions between them such as Ampère's law and Faraday's law. Such theory provides an adequate basis for the study of electrical power circuits and applications. However, communication systems are becoming an increasingly important part of modern civilisation. As communications become more sophisticated, they demand wide bandwidths and high-frequency circuits. Simple circuit theory then becomes inadequate to describe all the high-frequency phenomena.

The initial years of university degree courses include courses in elementary electromagnetism and field theory. For those who wish to specialise in electronics, it is then necessary to go on to a study of the more advanced electromagnetic theory to describe wave propagation on transmission lines and radio wave propagation. This book provides such a course of more advanced electromagnetic theory in its engineering applications. It is suitable for the second or third years of electrical engineering or electronics degree courses, and arises out of the author's experience of lecturing such courses over 25 years.

Communication systems today are the major users of high-frequency electrical signals and radio waves. So the electromagnetic theory introduced in this book is explained mainly in the context of communication systems applications over the complete frequency range from low-frequency a.c. up to optical frequencies. Chapter 1 starts with an introductory section followed by a statement of the essential fundamentals. Chapter 2 introduces the effect of the time of propagation of switching pulses on any long cable. Chapter 3 then goes on to the phase delay effects of an a.c. signal on a similar long cable. In Chapter 4 the elementary concepts of electromagnetic radiation are developed in the theory of the plane wave. In Chapter 5 the elementary concepts of confined guided waves are developed. Chapters 6 and 7 then give a detailed analysis of practical wave guiding systems in hollow metal waveguides, as used for microwave systems, and optical fibres respectively. Chapter 8 describes some antennas used to launch radio wave propagation and Chapter 9 describes some of the communication systems in which such radio wave propagation is used. Each section concludes with a summary of the theory developed in that section together with an example of the application of the theory and problems for the readers to attempt for themselves. The nine chapters are divided into 44 sections. There is a total of 60 worked examples and 80 problems provided at the ends of the sections.

Acknowledgements

There is a small overlap of subject matter between the contents of this book and the author's earlier book *Microwaves*. The new book forms a good introduction to the more specialised study given in the author's earlier book. The author is grateful to Pergamon Press Ltd for permission to copy some of the diagrams from *Microwaves*. The following diagrams have been reprinted with permission from

A. J. Baden Fuller, *Microwaves* (3rd edn), © 1990, Pergamon Press Ltd:

Figs. 3.6, 6.4, 6.6, 6.10, 6.11, 6.12, 6.18.

The following diagrams are similar to diagrams in *Microwaves*:

Figs. 2.2, 2.3, 3.2, 3.10, 5.2, 5.3, 5.5, 5.6, 5.7, 5.11, 5.16, 6.8, 6.15, 6.19.

Figure 7.9 is taken from P. Halley, *Fibre Optic Systems*, Copyright 1987. Reprinted by permission of John Wiley & Sons Ltd.

From a working life spent first in the microwave industry and then lecturing electromagnetic theory to university students it is impossible to identify all those from whom I have learned. I hope that lack of acknowledgement will not be taken to imply lack of gratitude. However, I should like to thank Professor P. Hammond for inspiring me with a love for electromagnetism and for his encouragement which inspired me to write my first book. I should like to thank Professor N. B. Jones, Head of the Department of Engineering at the University of Leicester, for his encouragement for me to continue writing, and I should like to thank Dr A. McCowen for encouraging me to contribute this book to the series of which he is an editor.

CHAPTER 1

Electromagnetic Fields

Aims: The aim of this chapter is to provide a summary of the important mathematical tools and electromagnetic field relationships necessary as an introduction to a study of applied time-varying electromagnetic fields.

1.1 INTRODUCTION

The introduction starts with a short historical summary.
 Electromagnetism is the name given to the study of electric and magnetic fields and their interactions. The elementary properties of electrostatic and magnetostatic fields may be largely studied in isolation. In 1600 Gilbert plotted the lines of magnetic force around a magnet. Later, other scientists investigated similar phenomena with electrostatic charges. Electric and magnetic field effects were thought to be completely independent until, after the invention of the electric battery, Oersted in 1820 discovered the magnetic effect of an electric current. Ampère's name is more usually associated with this effect because he gave a mathematical relationship for the phenomena. Then Faraday discovered electromagnetic induction in 1831, leading to the widespread design, manufacture and use of generators, motors and transformers. At this stage, apart from electromagnetic induction effects, it is still possible to study the properties of electric and magnetic fields in isolation. Such a study is the basis of most elementary courses in electromagnetism.
 In 1861 Maxwell published his theory which integrated electric and magnetic field effects and predicted the possibility of electromagnetic waves. Hertz proved the existence of these waves experimentally in 1888 but it was left to Marconi to realise their potential as a communication medium. He started his experiments in 1894 and in 1901 made the historic transmission of a radio message across the Atlantic. This book is a study of the mathematical properties of radio or electromagnetic waves across the whole spectrum. It is a study of electromagnetic waves, particularly in their engineering applications.
 In electromagnetic field theory, there is a distinct separation between the d.c. or low-frequency properties of the electric and magnetic fields, where

these can be treated largely independently, and the high-frequency properties, where the electric and magnetic fields interact in the wave. At low frequencies, the electric field is generated as the property of a distinct voltage and the magnetic field is generated by an electric current. At high frequencies, the voltage and current may be difficult to specify, but if they exist they are related by the properties of their associated electromagnetic fields. In terms of circuit properties, this relationship is specified by the characteristic impedance of a transmission line as described in Chapter 2.

In the present chapter, the basic mathematics essential to a study of electromagnetism is quoted, discussed and developed. First, all the field quantities are defined. Then the elementary electric and magnetic field relationships are quoted and discussed. The mathematical manipulation of electromagnetic field quantities is best performed in terms of the mathematical technique called vector analysis, which is introduced in the context of the electromagnetic fields. Then Maxwell's modification of the elementary electromagnetic field relationships is given, leading to that formulation of the relationships called Maxwell's equations. These equations are the basis of all the rest of the theory given in this book.

SUMMARY

Steady field or low-frequency effects in electrostatics can be described as the result of an electrical potential difference and those in magnetism as the result of an electrical current. The two fields are largely independent of each other.

Such studies are usually the subject of an elementary course on electromagnetism.

At high frequencies the electric and magnetic fields interact and the electric potential difference and electric current are related by the fields. A study of such relatively high-frequency fields is the subject of this book. The electromagnetic fields form radio waves, or light etc., depending on their frequency.

1.2 FIELD QUANTITIES

Electric charges and electric currents give rise to electric and magnetic fields. The basic mathematical relationships between the fields and the charges and currents are derived by deductive reasoning from experimental observations. These elementary relationships are usually explained in an elementary course in electromagnetism* and they will not be derived here. However, they are restated because they are the foundation of all the theory given in the rest of this book. First it is necessary to define all the field quantities and their usual units of measurement. As the fields, charges and currents all

---

*C. Christopoulos, *An Introduction to Applied Electromagnetism*, Wiley 1990.

exist distributed throughout the body of a medium, they are defined in terms of a space distribution. The notation and units of measurement of the electromagnetic field components are:

Electric field intensity	E	volt/metre
Electric flux density	D	coulomb/metre^2
Magnetic field intensity	H	ampère/metre
Magnetic flux density	B	tesla = weber/metre^2
Charge density	ρ	coulomb/metre^3
Current density	J	ampère/metre^2

The field intensity and the flux density are related by the permittivity and permeability of the medium. Therefore for the electric field

$$\mathbf{D} = \varepsilon \mathbf{E} \tag{1.1}$$

The permittivity ε is a dimensional constant which is also a property of the medium in which the fields exist. It is usually divided into two parts:

$$\varepsilon = \varepsilon_0 \varepsilon_r$$

The first, ε_0, is a dimensional constant called the *permittivity constant*, or sometimes the permittivity of free space. It is solely a function of the system of dimensions used. In SI units it is given by

$$\varepsilon_0 = 1/(36\pi \times 10^9) \quad \text{F/m}$$

The second, ε_r, is called the *relative permittivity*, which is dimensionless and makes allowance for the effect of the material relative to vacuum or free space. For a vacuum, eqn. (1.1) becomes

$$\mathbf{D} = \varepsilon_0 \mathbf{E}$$

so that for a vacuum, $\varepsilon_r = 1$. In any other material,

$$\mathbf{D} = \varepsilon_0 \varepsilon_r \mathbf{E} \tag{1.2}$$

For the magnetic field, the relationship is similar,

$$\mathbf{B} = \mu \mathbf{H} \tag{1.3}$$

The permeability μ is also a dimensional constant, which may also be divided into two parts:

$$\mu = \mu_0 \mu_r$$

The dimensional constant, μ_0, is called the *permeability constant* or the

4 ELECTROMAGNETIC FIELDS

permeability of free space. It is defined to be

$$\mu_0 = 4\pi \times 10^{-7} \quad \text{H/m}$$

The dimensionless constant μ_r is called the *relative permeability* and takes account of the effect of the medium on magnetic fields. For a vacuum $\mu_r = 1$ and for most non-magnetic materials $\mu_r \approx 1$; otherwise

$$\mathbf{B} = \mu_0 \mu_r \mathbf{H} \tag{1.4}$$

As will be shown in Chapter 4, the permittivity and permeability constants are related by the speed of light. The value of the permeability constant is defined by international agreement to be exactly $4\pi \times 10^{-7}$ H/m, whereas the value for the permittivity constant given above is only approximate and is the value obtained if the speed of light is taken to be 3×10^8 m/s.

A third important material relationship is given by Ohm's law, which in its distributed fields format is given by

$$\mathbf{J} = \sigma \mathbf{E} \tag{1.5}$$

An electric field intensity causes a current density to flow in a conductor. The material constant, σ, is the conductivity, measured in units of siemens/metre. It is the reciprocal of the resistivity.

The electric continuity equation states that the rate of change of charge inside a closed surface is equal to the total flow of current into that surface. Mathematically, it is expressed as

$$\int_{\text{volume}} \frac{\partial \rho}{\partial t} \, dv = -\int_{\text{area}} \mathbf{J} \cdot d\mathbf{s} \tag{1.6}$$

where d**s** is a small element of surface and the integral is taken over the surface of the volume containing the charge.

SUMMARY

Material properties,

$$\mathbf{D} = \varepsilon_0 \varepsilon_r \mathbf{E} \tag{1.2}$$

$$\mathbf{B} = \mu_0 \mu_r \mathbf{H} \tag{1.4}$$

$$\mathbf{J} = \sigma \mathbf{E} \tag{1.5}$$

The continuity equation,

$$\int_{\text{volume}} \frac{\partial \rho}{\partial t} \, dv = -\int_{\text{area}} \mathbf{J} \cdot d\mathbf{s} \tag{1.6}$$

ELECTROMAGNETIC FIELD RELATIONSHIPS

Example 1.1 A copper wire is carrying a current density of 5×10^6 A/m². Find the electric field intensity, which is the same as the potential gradient in the wire. The conductivity of copper is 5×10^7 S/m.

Answer Substituting values from the question into eqn. (1.5) gives

$$E = \frac{J}{\sigma} = \frac{5 \times 10^6}{5 \times 10^7} = 0.10 \text{ V/m}$$

PROBLEM

1.1 Most magnetic steels saturate at flux densities much above 1.5 T. A permanent magnet has mild steel pole pieces whose relative permeability may be taken to be 1000. The flux density in the air gap and in the pole pieces is 1.5 T. Find the value of the magnetic field intensity in the air gap and in the pole pieces.

[1.2×10^6 A/m, 1.2×10^3 A/m]

1.3 ELECTROMAGNETIC FIELD RELATIONSHIPS

The electromagnetic field equations are derived by deductive reasoning from the results of experiments. In each one, the field quantities are integrated throughout the region of interest so that these are called the integral form of the field equations. The primary field equations are as follows.

GAUSS'S LAW

The total flux flowing out of the surface of a body is equal to the net sum of the flux sources enclosed within the surface of that body. In electrostatics, the flux sources are the stored charge which gives rise to the electrostatic flux density. Therefore,

$$\int_{\text{surface}} \mathbf{D} \cdot \mathbf{ds} = \int_{\text{volume}} \rho \, dv \qquad (1.7)$$

The left-hand side of this equation is the normal component of the flux density integrated across the whole of the surface of the body, and the right-hand side is the electrostatic charge density integrated throughout the volume enclosed by the surface. In many practical situations, the right-hand side of the equation may be replaced by the sum of a number of point charges.

6 ELECTROMAGNETIC FIELDS

In magnetism there are no charges, so that the magnetic form of Gauss's law is

$$\int_{\text{surface}} \mathbf{B} \cdot \mathbf{ds} = 0 \qquad (1.8)$$

POTENTIAL

Electrostatic potential is defined from the work done in moving a unit charge in an electrostatic field. Then the electrostatic potential difference between the points a and b in a field is given by

$$V_a - V_b = -\int_b^a \mathbf{E} \cdot \mathbf{dl} \qquad (1.9)$$

The absolute potential is a value of potential measured with the zero of potential at infinity.

The electrostatic field is called a *conservative* field in that movement in the field around a closed path conserves energy and the potential difference between any two points is unique irrespective of the path traversed in obtaining the integral in eqn. (1.9). Taking the integral around a closed path is then a field statement of *Kirchhoff's voltage law*,

$$\oint \mathbf{E} \cdot \mathbf{dl} = 0 \qquad (1.10)$$

AMPÈRE'S LAW

Ampère's law is the magnetic field equivalent to eqn. (1.10) except that the magnetic field is not a conservative field in the presence of an electric current. The magnetic equivalent of potential is the integral of the magnetic field intensity around a closed path, which is equal to the total electric current flowing through the area bounded by that path, as shown in Figure 1.1. Mathematically, Ampère's law becomes

$$\oint \mathbf{H} \cdot \mathbf{dl} = \int_{\text{area}} \mathbf{J} \cdot \mathbf{ds} \qquad (1.11)$$

FARADAY'S LAW

If there is a changing magnetic flux density, then eqn. (1.10) no longer holds and the electrostatic field is not a conservative field. A potential difference is generated around a closed path when that closed path encompasses a changing magnetic flux. The potential difference is generated either by

ELETROMAGNETIC FIELD RELATIONSHIPS

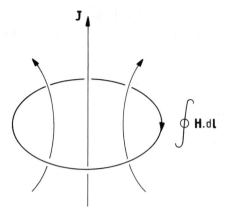

Figure 1.1 An illustration of Ampère's law. Integration around the closed loop is equal to the total current flowing through the loop.

changing the area enclosed by the path for a constant flux density or by the effect of a changing flux density enclosed by a constant path or both. Therefore

$$\oint \mathbf{E} \cdot \mathbf{dl} = -\frac{\partial}{\partial t} \int_{\text{area}} \mathbf{B} \cdot \mathbf{ds} \tag{1.12}$$

BOUNDARY CONDITIONS

At a boundary between two different media, the application of Gauss's law in the absence of any surface charge density at the boundary shows that the normal components of the flux density on each side of the boundary are equal. Therefore

$$D_{n1} = D_{n2} \tag{1.13}$$

and

$$B_{n1} = B_{n2} \tag{1.14}$$

Magnetic charges do not exist; therefore eqn. (1.14) is always true. In the electrostatic field however, charges can exist, and in the presence of a surface charge density at the boundary, eqn. (1.13) becomes

$$D_{n1} = D_{n2} + \rho_s \tag{1.15}$$

There is no electrostatic flux density inside a conductor, so that for a conductor, the electric field is always normal to the boundary of a conductor and the electric flux density at the surface has the same magnitude as the free charge density at the surface of the conductor.

The application of Kirchhoff's voltage law shows that the tangential components of the field intensity are the same on each side of the boundary. Therefore

$$E_{t1} = E_{t2}, \qquad (1.16)$$

and in the absence of any electric current,

$$H_{t1} = H_{t2} \qquad (1.17)$$

If medium 2 is a conductor, there will be a surface current density and eqn. (1.17) becomes

$$H_{t1} = H_{t2} + I_s \qquad (1.18)$$

where the direction of the current is also parallel to the surface but perpendicular to the magnetic field intensity. In a good conductor, the magnetic field intensity is negligible and eqn. (1.18) becomes

$$H_{t1} = I_s \qquad (1.19)$$

To summarise the effect of a boundary in a non-conductor with no surface charge: the normal components of the flux density and the tangential components of the field intensity are the same on each side of the boundary.

SUMMARY

Gauss's law:
$$\int_{\text{surface}} \mathbf{D} \cdot \mathbf{ds} = \int_{\text{volume}} \rho \, dv \qquad (1.7)$$

$$\int_{\text{surface}} \mathbf{B} \cdot \mathbf{ds} = 0 \qquad (1.8)$$

Potential:
$$V_a - V_b = -\int_b^a \mathbf{E} \cdot \mathbf{dl} \qquad (1.9)$$

Kirchhoff's voltage law:
$$\oint \mathbf{E} \cdot \mathbf{dl} = 0 \qquad (1.10)$$

Ampère's law:
$$\oint \mathbf{H} \cdot \mathbf{dl} = \int_{\text{area}} \mathbf{J} \cdot \mathbf{ds} \qquad (1.11)$$

Faraday's law:
$$\oint \mathbf{E} \cdot \mathbf{dl} = -\frac{\partial}{\partial t} \int_{\text{area}} \mathbf{B} \cdot \mathbf{ds} \qquad (1.12)$$

ELETROMAGNETIC FIELD RELATIONSHIPS

At the boundary between two non-conducting media having no free charges, the normal components of the flux density and the tangential components of the field intensity are the same on each side of the boundary.

Example 1.2 An isolated dielectric sphere of radius R has a relative permittivity ε_r. There is a point charge Q at the centre of the sphere. Find expressions for the absolute potential both inside and outside the dielectric sphere.

Answer First it is necessary to apply Gauss's law to the area surrounding the point charge. Choose a surface which is the surface of a sphere radius r centred on the point charge. By symmetry, the flux density vector is pointing radially outward and has the same magnitude everywhere on the surface of the sphere. Then the integral on the left-hand side of eqn. (1.7) is the value of the flux density multiplied by the area of the surface of the sphere, and the right-hand side is just the value of the point charge. Then

$$D \times 4\pi r^2 = Q \quad \text{and} \quad D = \frac{Q}{4\pi r^2}$$

everywhere, both inside and outside the dielectric sphere. This confirms the boundary statement that the normal components of the flux density are the same on both sides of the boundary. The electric field intensity is parallel to the flux density vector and is given by eqn. (1.2). It is different in the two media. Therefore

$$E = \frac{Q}{4\pi\varepsilon_0\varepsilon_r r^2} \qquad 0 < r < R$$

$$E = \frac{Q}{4\pi\varepsilon_0 r^2} \qquad R < r$$

The electrostatic potential in the field is determined by applying eqn. (1.9). If the integration is taken along a radial path, then both E and dl are parallel vectors. The absolute potential is a value of potential measured with the zero of potential at infinity. Then outside the dielectric sphere and substituting the value of E given above, eqn. (1.9) becomes

$$V - 0 = -\int_\infty^r E \cdot dl = -\int_\infty^r \frac{Q\,dr}{4\pi\varepsilon_0 r^2} = \frac{Q}{4\pi\varepsilon_0 r}, \qquad R < r$$

Inside the sphere it is necessary to perform the integration in two parts, therefore

$$V - 0 = -\int_\infty^r E \cdot dl = -\int_\infty^R \frac{Q\,dr}{4\pi\varepsilon_0 r^2} - \int_R^r \frac{Q\,dr}{4\pi\varepsilon_0\varepsilon_r r^2}, \qquad 0 < r < R$$

$$V = \frac{Q}{4\pi\varepsilon_0}\left(\frac{1}{R} + \frac{1}{\varepsilon_r r} - \frac{1}{\varepsilon_r R}\right), \qquad 0 < r < R$$

PROBLEMS

1.2 A coaxial cable has an inner conductor of radius a, a solid outer conductor of inner radius b and an insulator of relative permittivity ε_r. Given that the inner conductor carries a line charge density of $q\,C/m$, find the potential difference between the inner and outer conductors.

$$\left[V = \frac{q}{2\pi\varepsilon_0\varepsilon_r} \ln \frac{a}{b}\right]$$

1.3 Find the magnetic flux density a distance r from an isolated long straight wire carrying a current I.

$$\left[B = \frac{\mu_0 I}{2\pi r}\right]$$

1.4 VECTOR ANALYSIS

In the last section, the integral form of the electromagnetic field equations were given but in many instances it is more convenient to use the differential form of the same relationships. These are derived in this section. At the same time the opportunity is taken to introduce the vector differential operator. Use of the vector differential operator gives rise to that branch of mathematics called *vector analysis*.

GRADIENT

The definition of potential is given by eqn. (1.9) so that the electrostatic potential difference may be calculated from a knowledge of the electric field intensity. However, if the value of the potential in the field is known, then it is necessary to differentiate eqn. (1.9) in order to find the field intensity. Differentiating eqn. (1.9) gives

$$\mathbf{E} = -\frac{\partial V}{\partial \mathbf{l}} \tag{1.20}$$

where $\mathbf{E}$ is a vector which is tangential to the small element of line represented by $\delta\mathbf{l}$. In a system of rectangular coordinates, both sides of eqn. (1.20) may be separated into its three perpendicular components,

$$E_x = -\frac{\partial V}{\partial x}$$

$$E_y = -\frac{\partial V}{\partial y}$$

$$E_z = -\frac{\partial V}{\partial z}$$

where the subscripts denote the direction of that component of the vector. These equations are combined into one vector equation

$$\mathbf{E} = -\mathbf{U}_x \frac{\partial V}{\partial x} - \mathbf{U}_y \frac{\partial V}{\partial y} - \mathbf{U}_z \frac{\partial V}{\partial z} \quad (1.21)$$

where $\mathbf{U}_x$ is a unit vector in the x-direction and $\mathbf{U}_y$ and $\mathbf{U}_z$ are similarly defined. The differential operation involves directionality. Each component of the vector $\mathbf{E}$ is given by the gradient of the potential in the appropriate direction. Then collecting the total directional differential operation together into one vector operator gives

$$\mathbf{E} = -\left(\mathbf{U}_x \frac{\partial}{\partial x} + \mathbf{U}_y \frac{\partial}{\partial y} + \mathbf{U}_z \frac{\partial}{\partial z}\right) V = -\text{grad } V \quad (1.22)$$

The vector operator is represented by $\mathbf{\nabla}$ (del, sometimes called nabla):

$$\mathbf{\nabla} \equiv \mathbf{U}_x \frac{\partial}{\partial x} + \mathbf{U}_y \frac{\partial}{\partial y} + \mathbf{U}_z \frac{\partial}{\partial z}$$

Then eqn. (1.22) is written

$$\mathbf{E} = -\mathbf{\nabla} V = -\text{grad } V \quad (1.23)$$

The potential function V is a scalar so that the vector operator operating on a scalar gives a vector. Such an operation is called taking the gradient of V and can equally be represented mathematically by the mathematical function grad.

In a two-dimensional system, height contours on a map are a scalar potential quantity and the slope of the ground is the gradient of the height. The gradient is a vector quantity whose direction is perpendicular to the lines of constant height. Similarly in any potential field, the gradient is a vector which is perpendicular to the lines or surfaces of equipotential. Therefore in the electrostatic field, the electric field intensity is perpendicular to the surfaces of constant potential.

DIVERGENCE

Consider Gauss's law as given in eqn. (1.7). If this is applied to an elemental volume in three-dimensional space as shown in Figure 1.2, integrating over the complete cube, the left-hand side of the equation becomes

$$\int_{\text{surface}} \mathbf{D} \cdot \mathbf{ds} = \left(D_x + \frac{\partial D_x}{\partial x} \delta x\right) \delta y \, \delta z - D_x \delta y \, \delta z$$

$$+ \left(D_y + \frac{\partial D_y}{\partial y} \delta y\right) \delta x \, \delta z - D_y \delta x \, \delta z$$

$$+ \left(D_z + \frac{\partial D_z}{\partial z} \delta z\right) \delta x \, \delta y - D_z \delta x \, \delta y$$

12 ELECTROMAGNETIC FIELDS

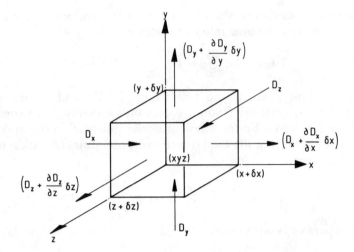

Figure 1.2 The derivation of divergence from Gauss's law, an elemental volume in three-dimensional space in the presence of a generalised D field.

The right-hand side of eqn. (1.7) gives

$$\int_{\text{volume}} \rho \, dv = \rho \, \delta x \, \delta y \, \delta z$$

assuming that the elemental volume is sufficiently small for changes in the charge density ρ to be negligible. Therefore, putting the two sides of the equation together and simplifying gives

$$\frac{\partial D_x}{\partial x} + \frac{\partial D_y}{\partial y} + \frac{\partial D_z}{\partial z} = \rho \qquad (1.24)$$

The vector operation given in eqn. (1.24) is defined by

$$\text{div}\,\mathbf{D} = \rho \qquad (1.25)$$

or using the vector operator,

$$\text{div}\,\mathbf{D} = \boldsymbol{\nabla} \cdot \mathbf{D} = \rho \qquad (1.26)$$

where

$$\left(\mathbf{U}_x \frac{\partial}{\partial x} + \mathbf{U}_y \frac{\partial}{\partial y} + \mathbf{U}_z \frac{\partial}{\partial z}\right) \cdot (\mathbf{U}_x D_x + \mathbf{U}_y D_y + \mathbf{U}_z D_z) = \frac{\partial D_x}{\partial x} + \frac{\partial D_y}{\partial y} + \frac{\partial D_z}{\partial z}$$

Then eqn. (1.26) is another statement of Gauss's law for the electrostatic field. The similar statement of Gauss's law for the magnetic field is

$$\text{div}\,\mathbf{B} = \boldsymbol{\nabla} \cdot \mathbf{B} = 0 \qquad (1.27)$$

VECTOR ANALYSIS

The divergence operation is the same as the scalar product of the vector operator with the vector on which it operates.

LAPLACIAN

Consider the two electrostatic field equations that have been derived so far, eqns. (1.23) and (1.26). Substituting the material properties, eqn. (1.1), into eqn. (1.26) gives

$$\text{div } \mathbf{E} = \nabla \cdot \mathbf{E} = \frac{\rho}{\varepsilon_0}$$

Substituting for **E** from eqn. (1.23) gives

$$-\text{div grad } V = -\nabla \cdot \nabla V = \frac{\rho}{\varepsilon_0} \qquad (1.28)$$

$\nabla \cdot \nabla$ is written as ∇^2, and as an operator becomes

$$\nabla^2 = \frac{\partial^2}{\partial x^2} + \frac{\partial^2}{\partial y^2} + \frac{\partial^2}{\partial z^2}$$

Then eqn. (1.28) becomes

$$\nabla^2 V = \frac{\partial^2 V}{\partial x^2} + \frac{\partial^2 V}{\partial y^2} + \frac{\partial^2 V}{\partial z^2} = -\frac{\rho}{\varepsilon_0} \qquad (1.29)$$

Equation (1.29) is called *Poisson's equation* and occurs in a number of engineering field systems, not just in electrostatics. If we are dealing with electric fields apart from a conductor or any stored charge, then $\rho = 0$ and eqn. (1.29) becomes

$$\nabla^2 V = 0 \qquad (1.30)$$

which is called *Laplace's equation*. The operator ∇^2 is called *the Laplacian* which can operate on a vector as well as on a scalar. Therefore

$$\nabla^2 \mathbf{E} = \frac{\partial^2 \mathbf{E}}{\partial x^2} + \frac{\partial^2 \mathbf{E}}{\partial y^2} + \frac{\partial^2 \mathbf{E}}{\partial z^2}$$

If

$$\nabla^2 \mathbf{E} = k^2 \mathbf{E}$$

then

$$\frac{\partial^2 E_x}{\partial x^2} + \frac{\partial^2 E_x}{\partial y^2} + \frac{\partial^2 E_x}{\partial z^2} = k^2 E_x$$

14 ELECTROMAGNETIC FIELDS

and two similar equations in E_y and E_z. The del operator is just a shorthand but it is an operator that obeys many of the rules of vector algebra.

GAUSS'S THEOREM

Gauss's theorem (not to be confused with Gauss's law) can be derived from what we have just learnt about Gauss's law. Substituting from eqn. (1.26) into eqn. (1.7) gives

$$\int_{\text{surface}} \mathbf{D} \cdot \mathbf{ds} = \int_{\text{volume}} \rho \, dv = \int_{\text{volume}} \text{div} \, \mathbf{D} \, dv \qquad (1.31)$$

The relationship given in eqn. (1.31) is that the integral of the normal component of a vector field quantity over a surface is equal to the integral over the volume enclosed by that surface of the divergence of the same vector. This is true for any vector and is not just a property of the electrostatic field. Therefore, if $\mathbf{F}$ is any vector, Gauss's theorem or the divergence theorem states that

$$\int_{\text{surface}} \mathbf{F} \cdot \mathbf{ds} = \int_{\text{volume}} \text{div} \, \mathbf{F} \, dv \qquad (1.32)$$

CURL

In Ampère's law and Faraday's law, the m.m.f. and the e.m.f. around a closed path are given by the left-hand side of eqns. (1.11) and (1.12) respectively:

$$\text{m.m.f.} = -\oint \mathbf{H} \cdot \mathbf{dl} \qquad (1.33)$$

$$\text{e.m.f.} = -\oint \mathbf{E} \cdot \mathbf{dl} \qquad (1.34)$$

and these are each equal to the total flux threading the loop. In order to consider the differential form of these relationships, take an incremental loop in the x–y plane as shown in Figure 1.3. Therefore

$$\text{e.m.f.} = \oint \mathbf{E} \cdot \mathbf{dl} = E_x \delta x - \left(E_x + \frac{\partial E_x}{\partial y} \delta y \right) \delta x$$

$$- E_y \delta y + \left(E_y + \frac{\partial E_y}{\partial x} \delta x \right) \delta y$$

$$= \left(\frac{\partial E_y}{\partial x} - \frac{\partial E_x}{\partial y} \right) \delta x \, \delta y$$

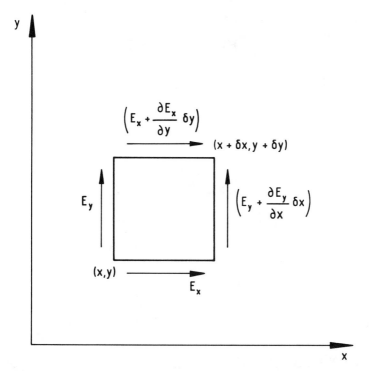

Figure 1.3 Illustrating curl, an incremental loop in the x–y plane.

This e.m.f. is equal to the flux density perpendicular to the plane of the loop multiplied by the area of the loop, assuming that the dimensions of the loop are so small that any variation in the flux density across the area of the loop is negligible. Therefore

$$F_z \delta x\, \delta y = \left(\frac{\partial E_y}{\partial x} - \frac{\partial E_x}{\partial y}\right) \delta x\, \delta y$$

or cancelling out the area of the loop,

$$F_z = \frac{\partial E_y}{\partial x} - \frac{\partial E_x}{\partial y}$$

Similarly it can be shown that

$$F_x = \frac{\partial E_z}{\partial y} - \frac{\partial E_y}{\partial z}$$

$$F_y = \frac{\partial E_x}{\partial z} - \frac{\partial E_z}{\partial x}$$

This operation on **E** is called curl and is the same as vector multiplication by the vector operator ∇. Therefore

$$F = \text{curl}\,E = \nabla \times E \tag{1.35}$$

or

$$\left.\begin{aligned} F_x &= \frac{\partial E_z}{\partial y} - \frac{\partial E_y}{\partial z} \\ F_y &= \frac{\partial E_x}{\partial z} - \frac{\partial E_z}{\partial x} \\ F_z &= \frac{\partial E_y}{\partial x} - \frac{\partial E_x}{\partial y} \end{aligned}\right\} \tag{1.36}$$

As with vector multiplication, eqn. (1.36) can be written as a determinant:

$$\text{curl}\,E = \begin{vmatrix} \mathbf{U}_x & \mathbf{U}_y & \mathbf{U}_z \\ \dfrac{\partial}{\partial x} & \dfrac{\partial}{\partial y} & \dfrac{\partial}{\partial z} \\ E_x & E_y & E_z \end{vmatrix} \tag{1.37}$$

STOKES'S THEOREM

Stokes's theorem is the curl analogue of Gauss's theorem. It has already been shown, in the development of eqn. (1.35), that

$$\oint \mathbf{E} \cdot \mathbf{dl} = \text{total flux threading the loop} = \int_{\text{area}} \mathbf{F} \cdot \mathbf{ds}$$

Then substituting for $\mathbf{F}$ from eqn. (1.35),

$$\oint \mathbf{E} \cdot \mathbf{dl} = \int_{\text{area}} \text{curl}\,\mathbf{E} \cdot \mathbf{ds} \tag{1.38}$$

Equation (1.38) is a mathematical statement of Stokes's theorem, that for any vector quantity the line integral of the vector around any closed loop is equal to the integral of the curl of that vector across any area bounded by that loop. Again, it is a mathematical identity and not just a property of the electrostatic field.

Applying Stokes's theorem to Ampère's law, eqn. (1.11), gives

$$\int \mathbf{H} \cdot \mathbf{dl} = \int_{\text{area}} \text{curl}\,\mathbf{H} \cdot \mathbf{ds} = \int_{\text{area}} \mathbf{J} \cdot \mathbf{ds} \tag{1.39}$$

Therefore, since the two area integrals are the same,

$$\text{curl}\,\mathbf{H} = \nabla \times \mathbf{H} = \mathbf{J} \tag{1.40}$$

which is the differential form of Ampère's law. Similarly if Stokes's law is applied to Faraday's law, eqn. (1.12),

$$\oint \mathbf{E} \cdot \mathbf{dl} = \int_{\text{area}} \text{curl}\, \mathbf{E} \cdot \mathbf{ds} = -\frac{\partial}{\partial t} \int_{\text{area}} \mathbf{B} \cdot \mathbf{ds} \qquad (1.41)$$

Provided that the boundaries of the area of integration do not move with time, eqn. (1.41) may be written

$$\int_{\text{area}} \text{curl}\, \mathbf{E} \cdot \mathbf{ds} = \int_{\text{area}} -\frac{\partial \mathbf{B}}{\partial t} \cdot \mathbf{ds}$$

Therefore

$$\text{curl}\, \mathbf{E} = \nabla \times \mathbf{E} = -\frac{\partial \mathbf{B}}{\partial t} \qquad (1.42)$$

which is the differential form of Faraday's law.

VECTOR OPERATOR IDENTITIES

There are a number of useful identities which exist for vector operators. In eqn. (1.28) it has already been stated that

$$\text{div grad}\, V = \nabla \cdot \nabla V = \nabla^2 V \qquad (1.43)$$

however, the relationship has not yet been proved. Writing out the relationship in eqn. (1.43) in its component parts in rectangular coordinates gives

$$\left(\mathbf{U}_x \frac{\partial}{\partial x} + \mathbf{U}_y \frac{\partial}{\partial y} + \mathbf{U}_z \frac{\partial}{\partial z} \right) \cdot \left(\mathbf{U}_x \frac{\partial V}{\partial x} + \mathbf{U}_y \frac{\partial V}{\partial y} + \mathbf{U}_z \frac{\partial V}{\partial z} \right)$$

$$= \frac{\partial^2 V}{\partial x^2} + \frac{\partial^2 V}{\partial y^2} + \frac{\partial^2 V}{\partial z^2} = \nabla^2 V$$

hence verifying eqn. (1.43). Similarly, it can be shown that

$$\text{curl grad}\, V = \nabla \times \nabla V = 0 \qquad (1.44)$$

$$\text{div curl}\, \mathbf{E} = \nabla \cdot \nabla \times \mathbf{E} = 0 \qquad (1.45)$$

Another identity which will be used later in the solution of Maxwell's equations is

$$\text{curl curl}\, \mathbf{E} = \nabla \times (\nabla \times \mathbf{E}) = \text{grad div}\, \mathbf{E} - \nabla^2 \mathbf{E} = \nabla(\nabla \cdot \mathbf{E}) - \nabla^2 \mathbf{E} \qquad (1.46)$$

Now to relate the vector operator to the properties of certain types of field. Consider the fields shown in Figure 1.4. Figure 1.4(a) shows a uniform field,

18 ELECTROMAGNETIC FIELDS

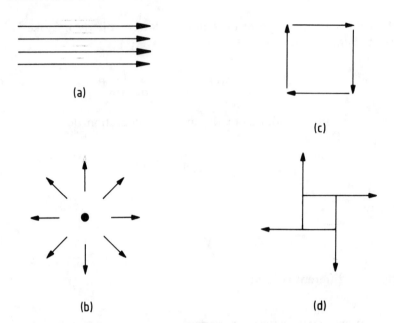

Figure 1.4 Illustrating various types of field: (a) a uniform field has a gradient but no divergence or curl; (b) a radial field has a gradient and divergence but no curl; (c) a solenoidal field has curl but no divergence; (d) a general field has both divergence and curl.

which has no divergence or curl but can have a gradient. This is also an irrotational conservative field that can be described by a scalar potential. Figure 1.4(b) shows the field radiating from a point source or a line source. This is a field that has a gradient and a divergence but not curl. It is an irrotational field and can also be described by a scalar potential. Figure 1.4(c) shows the field of a vortex. Such a field has curl but no divergence. It is often called a solenoidal field. The magnetic field is such a field. Figure 1.4(d) shows a completely general field that has both divergence and curl. It is neither conservative nor irrotational.

To return to the vector operator identities, from eqn. (1.23)

$$\mathbf{E} = -\text{grad } V$$

but this equation describes a field that is necessarily irrotational from eqn. (1.10) but expressed in differential form,

$$\text{curl } \mathbf{E} = 0$$

Therefore

$$\text{curl grad } V = 0 \tag{1.44}$$

This is also a mathematical basis for the statement that an irrotational conservative field can be completely described by a scalar potential, in the electrostatic field, the voltage V. Consider the magnetic field, and let it be described in terms of another vector $\mathbf{A}$ such that

$$\mathbf{B} = \operatorname{curl} \mathbf{A} \tag{1.47}$$

Substituting into eqn. (1.27) then shows that

$$\operatorname{div} \operatorname{curl} \mathbf{A} = 0$$

which is a restatement of eqn. (1.45). The properties of a solenoidal field are that it has curl but no divergence and the magnetic field is such a field. A solenoidal field can also be expressed as the curl of another vector as shown by eqn. (1.47). In this case, $\mathbf{A}$ is called the vector potential of the field.

SUMMARY

The vector operator

$$\boldsymbol{\nabla} = \mathbf{U}_x \frac{\partial}{\partial x} + \mathbf{U}_y \frac{\partial}{\partial y} + \mathbf{U}_z \frac{\partial}{\partial z}$$

Electrostatic potential is given by

$$\mathbf{E} = -\operatorname{grad} V = -\boldsymbol{\nabla} V \tag{1.23}$$

Gauss's law,

$$\operatorname{div} \mathbf{D} = \boldsymbol{\nabla} \cdot \mathbf{D} = \rho \tag{1.26}$$

$$\operatorname{div} \mathbf{B} = \boldsymbol{\nabla} \cdot \mathbf{B} = 0 \tag{1.27}$$

Poisson's equation

$$\nabla^2 V = -\frac{\rho}{\varepsilon_0} \tag{1.29}$$

Laplace's equation

$$\nabla^2 V = 0 \tag{1.30}$$

Gauss's theorem

$$\int_{\text{surface}} \mathbf{F} \cdot \mathbf{ds} = \int_{\text{volume}} \operatorname{div} \mathbf{F} \, dv \tag{1.32}$$

ELECTROMAGNETIC FIELDS

$$\text{curl } \mathbf{E} = \boldsymbol{\nabla} \times \mathbf{E} = \begin{vmatrix} \mathbf{U}_x & \mathbf{U}_y & \mathbf{U}_z \\ \dfrac{\partial}{\partial x} & \dfrac{\partial}{\partial y} & \dfrac{\partial}{\partial z} \\ E_x & E_y & E_z \end{vmatrix} \qquad (1.37)$$

Stokes's theorem

$$\oint \mathbf{E} \cdot d\mathbf{l} = \int_{\text{area}} \text{curl } \mathbf{E} \cdot d\mathbf{s} \qquad (1.38)$$

Ampère's law

$$\text{curl } \mathbf{H} = \boldsymbol{\nabla} \times \mathbf{H} = \mathbf{J} \qquad (1.40)$$

Faraday's law

$$\text{curl } \mathbf{E} = \boldsymbol{\nabla} \times \mathbf{E} = -\frac{\partial \mathbf{B}}{\partial t} \qquad (1.42)$$

Vector operator identities

$$\text{curl grad } V = \boldsymbol{\nabla} \times \boldsymbol{\nabla} V = 0 \qquad (1.44)$$

$$\text{div curl } \mathbf{E} = \boldsymbol{\nabla} \cdot \boldsymbol{\nabla} \times \mathbf{E} = 0 \qquad (1.45)$$

$$\text{curl curl } \mathbf{E} = \boldsymbol{\nabla} \times (\boldsymbol{\nabla} \times \mathbf{E}) = \text{grad div } \mathbf{E} - \nabla^2 \mathbf{E} = \boldsymbol{\nabla}(\boldsymbol{\nabla} \cdot \mathbf{E}) - \nabla^2 \mathbf{E} \quad (1.46)$$

The vector potential is defined by

$$\mathbf{B} = \text{curl } \mathbf{A} \qquad (1.47)$$

Example 1.3 A uniform electrostatic field is described by a uniform x-directed electric field intensity of value c V/m. Show that the divergence and curl of the field intensity are zero and find the scalar potential needed to describe the field.

Answer The field is

$$\mathbf{E} = \mathbf{U}_x c$$

Therefore

$$\text{div } \mathbf{E} = \frac{\partial(c)}{\partial x} = 0$$

and

$$\text{curl } \mathbf{E} = \mathbf{U}_y \frac{\partial(c)}{\partial z} - \mathbf{U}_z \frac{\partial(c)}{\partial y} = 0$$

The electric field intensity is the gradient of the voltage,

$$\mathbf{E} = -\text{grad}\, V$$

Therefore

$$c = -\frac{\partial V}{\partial x}$$

$$\frac{\partial V}{\partial y} = \frac{\partial V}{\partial z} = 0$$

Therefore, integrating gives

$$V = -cx + \text{const}$$

Example 1.4 By expanding the functions into three perpendicular components parallel to the axes of the rectangular coordinate system, verify the identity given in eqn. (1.46).

Answer To find the expansion of curl curl **E**, let curl **E** = **F**. Then

$$\text{curl curl}\, \mathbf{E} = \text{curl}\, \mathbf{F} = \mathbf{G}$$

Consider the x-directed component of **G**,

$$G_x = \frac{\partial F_z}{\partial y} - \frac{\partial F_y}{\partial z}$$

and

$$F_y = \frac{\partial E_x}{\partial z} - \frac{\partial E_z}{\partial x}$$

$$F_z = \frac{\partial E_y}{\partial x} - \frac{\partial E_x}{\partial y}$$

Substituting into the expression for G_x,

$$G_x = \frac{\partial^2 E_y}{\partial y \partial x} - \frac{\partial^2 E_x}{\partial y^2} - \frac{\partial^2 E_x}{\partial z^2} + \frac{\partial^2 E_z}{\partial z \partial x} + \left(\frac{\partial^2 E_x}{x^2} - \frac{\partial^2 E_x}{x^2}\right)$$

The additional terms in brackets are equal to zero and are added so that the terms may be collected together to give the x-directed component of the right-hand side of eqn. (1.46):

$$G_x = \frac{\partial}{\partial x}\left(\frac{\partial E_x}{\partial x} + \frac{\partial E_y}{\partial y} + \frac{\partial E_z}{\partial z}\right) - \left(\frac{\partial^2 E_x}{\partial x^2} + \frac{\partial^2 E_x}{\partial y^2} + \frac{\partial^2 E_x}{\partial z^2}\right)$$

There are similar expressions for the y and z-directed components. Therefore

$$\text{curl curl}\, \mathbf{E} = \text{grad}\,(\text{div}\, \mathbf{E}) - \nabla^2 \mathbf{E} \qquad (1.46)$$

PROBLEMS

grad:
1.4 Find grad V of the scalar fields,
   (a) $V = x^2 + 3y^2 - 2z^2$

   $[2x\mathbf{U}_x + 6y\mathbf{U}_y - 4z\mathbf{U}_z]$

   (b) $V = \frac{1}{2}\ln(x^2 + y^2)$

   $[(x\mathbf{U}_x + y\mathbf{U}_y)/(x^2 + y^2)]$

   (c) $V = 3x^2y - y^3z^2$ at the point $(1, -2, -1)$

   $[-12\mathbf{U}_x - 9\mathbf{U}_y - 16\mathbf{U}_z]$

1.5 Find the unit normal vector to the given surface at the point $P(x, y, z)$
   (a) $x^2 + y^2 + 2z^2 = 26$ P is $(2, 2, 3)$

   $[(4\mathbf{U}_x + 4\mathbf{U}_y + 12\mathbf{U}_z)/\sqrt{176}]$

   (b) $z^2 = x^2 + y^2$ P is $(3, 4, 5)$

   $[(3\mathbf{U}_x + 4\mathbf{U}_y - \mathbf{U}_z)/5\sqrt{2}]$

   (c) $z = y^4$ P is $(2, 1, 1)$

   $[(4\mathbf{U}_x - \mathbf{U}_y)/\sqrt{17}]$

   (d) $x^2y + 2xz = 4$ P is $(2, -2, 3)$

   $[(-\mathbf{U}_x + 2\mathbf{U}_y + 2\mathbf{U}_z)/3]$

div:
1.6 Find div $\mathbf{A}$ for the following vector functions
   (a) $\mathbf{A} = 2y\mathbf{U}_x + z\mathbf{U}_y + xy\mathbf{U}_z$
   (b) $\mathbf{A} = (x\mathbf{U}_x + y\mathbf{U}_y)/(x^2 + y^2)$
   (c) $\mathbf{A} = xyz(x\mathbf{U}_x + y\mathbf{U}_y + z\mathbf{U}_z)$

   $[0; 0; 6xyz]$

1.7 Given that $V = 2x^3y^2z^4$, find $\nabla^2 V$ at $(1, -1, 1)$

   $[40]$

curl:
1.8 Find curl $\mathbf{A}$ for the following vector functions
   (a) $\mathbf{A} = yz\mathbf{U}_x + zx\mathbf{U}_y + xy\mathbf{U}_z$

   $[0]$

   (b) $\mathbf{A} = y^2\mathbf{U}_x + z^2\mathbf{U}_y + x^2\mathbf{U}_z$

   $[-2z\mathbf{U}_x - 2x\mathbf{U}_y - 2y\mathbf{U}_z]$

   (c) $\mathbf{A} = xz^3\mathbf{U}_x - 2x^2yz\mathbf{U}_y + 2yz^4\mathbf{U}_z$ at the point $(1, -1, 1)$.

   $[3\mathbf{U}_y + 4\mathbf{U}_z]$

Mixed operators:
1.9 When $\mathbf{A} = 2yz\mathbf{U}_x - x^2y\mathbf{U}_y + xz^2\mathbf{U}_z$ and $V = 2x^2yz^3$ find $(\mathbf{A} \cdot \nabla)V$; $\mathbf{A} \cdot \nabla V$; $(\mathbf{A} \times \nabla)V$; $\mathbf{A} \times \nabla V$

[(i) & (ii) $8xy^2z^4 - 2x^4yz^3 + 6x^3yz^4$;
(iii) & (iv) $-(6x^4y^2z^2 + 2x^3z^3)\mathbf{U}_x - (12x^2y^2z^3 - 4x^3yz^5)\mathbf{U}_y + (4x^2yz^4 + 4x^3y^2z^3)\mathbf{U}_z$]

1.10 By expanding into the rectangular components of the vectors, prove some of the following identities:

$$\text{grad}(uv) = u \,\text{grad}\, v + v \,\text{grad}\, u$$

$$\text{div}(V\mathbf{A}) = \text{grad}\, V \cdot \mathbf{A} + V \,\text{div}\, \mathbf{A}$$

$$\text{curl}(V\mathbf{A}) = \text{grad}\, V \times \mathbf{A} + V \,\text{curl}\, \mathbf{A}$$

$$\text{grad}(\mathbf{A} \cdot \mathbf{B}) = (\mathbf{B} \cdot \nabla)\mathbf{A} + (\mathbf{A} \cdot \nabla)\mathbf{B} + \mathbf{B} \times \text{curl}\, \mathbf{A} + \mathbf{A} \times \text{curl}\, \mathbf{B}$$

$$\text{div}(\mathbf{A} \times \mathbf{B}) = \mathbf{B} \cdot \text{curl}\, \mathbf{A} - \mathbf{A} \cdot \text{curl}\, \mathbf{B}$$

$$\text{curl}(\mathbf{A} \times \mathbf{B}) = (\mathbf{B} \cdot \nabla)\mathbf{A} - (\mathbf{A} \cdot \nabla)\mathbf{B} + \mathbf{A}\,\text{div}\, \mathbf{B} - \mathbf{B}\,\text{div}\, \mathbf{A}$$

1.5 MAXWELL'S EQUATIONS

In the previous section a system of electromagnetic field equations has been developed in their differential form. However, they are only true and complete for static fields and slowly changing field quantities. As given in eqn. (1.40), Ampère's law is incomplete. Consider eqn. (1.40) again:

$$\text{curl}\, \mathbf{H} = \mathbf{J} \qquad (1.40)$$

Taking the divergence of each side of eqn. (1.40) gives

$$\text{div}\,\text{curl}\, \mathbf{H} = \text{div}\, \mathbf{J} \qquad (1.48)$$

But eqn. (1.45) shows that the left-hand side of eqn. (1.48) is identically equal to zero, therefore eqn. (1.48) becomes

$$\text{div}\, \mathbf{J} = 0 \qquad (1.49)$$

However, applying Gauss's theorem, eqn. (1.32), to the electrical continuity equation, eqn. (1.6), gives the differential form of that equation,

$$\text{div}\, \mathbf{J} = -\frac{\partial \rho}{\partial t} \qquad (1.50)$$

which is clearly at variance with eqn. (1.49). The left-hand side of eqn. (1.48) is identically equal to zero; therefore it is necessary to add another term to the right-hand side of that equation to make it equal to zero. Substituting from eqn. (1.50),

$$\text{div } \mathbf{J} + \frac{\partial \rho}{\partial t} = 0 = \text{div curl } \mathbf{H}$$

Substituting for ρ from eqn. (1.26) gives

$$\text{div } \mathbf{J} + \frac{\partial}{\partial t}(\text{div } \mathbf{D}) = \text{div curl } \mathbf{H} \tag{1.51}$$

Integrating eqn. (1.51) gives

$$\text{curl } \mathbf{H} = \mathbf{J} + \frac{\partial \mathbf{D}}{\partial t} \tag{1.52}$$

which shows that eqn. (1.40) is true only for static fields or when the field quantities are varying slowly.

However, there is also a simple physical justification for the extra term that has had to be added to the right-hand side of eqn. (1.52). Consider an a.c. circuit containing a parallel plate capacitor as shown in Figure 1.5. It is easy to show that, if the integral form of Ampère's law, eqn. (1.11), is applied to a circular path enclosing the wire,

$$H = \frac{I}{2\pi r}$$

However, without disturbing the path along which $\oint \mathbf{H} \cdot \mathbf{dl}$ is measured it is simple to move the surface across which the current is integrated to pass between the plates of the parallel plate capacitor so that the integral on the right-hand side of eqn. (1.11) is zero. Another term needs to be added to the right-hand side of eqn. (1.11) to allow for whatever is happening between

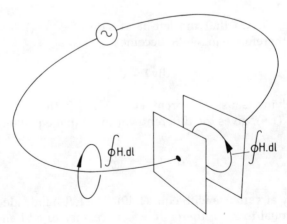

Figure 1.5 Illustrating Maxwell's difficulty with Ampère's law.

the plates of the capacitor. The circuit containing the capacitor must have a continuum of current flowing round it. Since no conduction current is flowing between the plates of the capacitor, the two halves of the circuit are linked by the electric flux between the plates of the capacitor. Then the current is equal to the rate of change of charge on the plates of the capacitor, which in its turn is equal to the rate of change of flux density between the plates of the capacitor. Therefore eqn. (1.11) is also incomplete and needs an additional term to be added to its right-hand side.

$$\oint \mathbf{H} \cdot \mathbf{dl} = \int_{\text{area}} \left(\mathbf{J} + \frac{\partial \mathbf{D}}{\partial t} \right) \cdot \mathbf{ds} \qquad (1.53)$$

The changing flux density between the plates of the capacitor is termed the *displacement current* in distinction to the *conduction current* flowing in the wires of the rest of the circuit. The four electromagnetic field equations are now complete and are gathered together here in their differential form. They are also written using the mathematical operator del:

$$\nabla \cdot \mathbf{D} = \rho \qquad (1.26)$$

$$\nabla \cdot \mathbf{B} = 0 \qquad (1.27)$$

$$\nabla \times \mathbf{E} = -\frac{\partial \mathbf{B}}{\partial t} \qquad (1.42)$$

$$\nabla \times \mathbf{H} = \mathbf{J} + \frac{\partial \mathbf{D}}{\partial t} \qquad (1.52)$$

Maxwell first formulated the electromagnetic field relationships into the form given here and these equations are called by his name.

The energy density stored in an electrostatic field is given by

$$w_e = \tfrac{1}{2} \mathbf{D} \cdot \mathbf{E} \qquad (1.54)$$

Similarly, the stored magnetic energy density is given by

$$w_m = \tfrac{1}{2} \mathbf{B} \cdot \mathbf{H} \qquad (1.55)$$

For most situations, the flux density and the field intensity are parallel vectors and the permittivity and permeability are constants so that the total stored energy due to the electromagnetic field quantities is

$$w_e + w_m = \tfrac{1}{2} DE + \tfrac{1}{2} BH = \tfrac{1}{2} \varepsilon E^2 + \tfrac{1}{2} \mu H^2 \qquad (1.56)$$

Then the rate of change of energy stored in the electromagnetic field is given by

$$\frac{\partial}{\partial t}(w_e + w_m) = \varepsilon E \frac{\partial E}{\partial t} + \mu H \frac{\partial H}{\partial t} \qquad (1.57)$$

For a completely general electromagnetic situation, it is necessary to add a further term to eqn. (1.57) that represents the electrical energy dissipated as heat due to any conduction current in the material. Therefore the total change of electromagnetic energy is given by

$$p = \varepsilon E \frac{\partial E}{\partial t} + \mu H \frac{\partial H}{\partial t} + \mathbf{E} \cdot \mathbf{J}$$

$$= \mathbf{E} \cdot \frac{\partial \mathbf{D}}{\partial t} + \mathbf{H} \cdot \frac{\partial \mathbf{B}}{\partial t} + \mathbf{E} \cdot \mathbf{J} = \mathbf{E} \cdot \left(\mathbf{J} + \frac{\partial \mathbf{D}}{\partial t} \right) + \mathbf{H} \cdot \frac{\partial \mathbf{B}}{\partial t}$$

Substituting from eqns. (1.42) and (1.52) gives

$$p = \mathbf{E} \cdot \operatorname{curl} \mathbf{H} - \mathbf{H} \cdot \operatorname{curl} \mathbf{E} \tag{1.58}$$

The right-hand side of eqn. (1.58) is the expansion of div $(\mathbf{H} \times \mathbf{E})$ as shown in Problem 1.10; therefore eqn. (1.58) becomes

$$p = -\operatorname{div}(\mathbf{E} \times \mathbf{H}). \tag{1.59}$$

The expression given in eqn. (1.59) describes the rate of change of electromagnetic stored energy density. To give practical results, this expression when integrated throughout any volume gives the net power flow into that volume. Therefore

$$P = \int_{\text{volume}} p \, dv = -\int_{\text{volume}} \operatorname{div}(\mathbf{E} \times \mathbf{H}) \, dv \tag{1.60}$$

Applying Gauss's theorem, eqn. (1.32), to the right-hand side of eqn. (1.60) gives

$$P = -\int_{\text{surface}} (\mathbf{E} \times \mathbf{H}) \cdot d\mathbf{s} \tag{1.61}$$

The vector product in eqns (1.59)–(1.61) is called the *Poynting vector* and is defined by

$$\mathbf{S} = \mathbf{E} \times \mathbf{H} \tag{1.62}$$

The Poynting vector may be loosely described as the power flow density in the electromagnetic field. Strictly speaking, it is the integral of the Poynting vector over a closed surface which is equal to the total electromagnetic energy flow through that surface. The negative sign occurs in eqn. (1.61) because the integral in the right-hand side of that equation describes power flow *out* of the surface, which leads to a *reduction* of stored energy.

Therefore, the power flow through the surface is equal to the time average of the integral of the normal component of the Poynting vector over that surface.

For electromagnetic fields in a non-conducting medium, there is no dissipation of power in the medium, there is no resistive power loss in the medium and the integral of the Poynting vector over any closed volume is equal to the power flow through the surface enclosing the volume. For an electric circuit enclosed inside a conducting screen, the electric field intensity in the conductor is zero, and the Poynting vector in the conductor is zero. Then the electrical power flow out of the termination is equal to the integral of the Poynting vector across the cross-section of the cable comprising the termination. Consequently, the power flow down a closed conductor such as a coaxial cable is assumed to be equal to the integral of the Poynting vector across the cross-section of the cable.

At low frequencies, it is easier to determine the power flow as the product of current and voltage but at high frequencies, when the electromagnetic radiation is often enclosed inside a hollow metal pipe, current and voltage are not easily identifiable quantities and the Poynting vector is used to calculate the power flow in the circuit. As shown in Example 1.5, the integral of the Poynting vector across the cross-section of a closed electric cable correctly gives the power flow even in a circuit carrying d.c.

SUMMARY

Maxwell's equations are

$$\nabla \cdot \mathbf{D} = \rho \quad (1.26)$$

$$\nabla \cdot \mathbf{B} = 0 \quad (1.27)$$

$$\nabla \times \mathbf{E} = -\frac{\partial \mathbf{B}}{\partial t} \quad (1.42)$$

$$\nabla \times \mathbf{H} = \mathbf{J} + \frac{\partial \mathbf{D}}{\partial t} \quad (1.52)$$

The Poynting vector is

$$\mathbf{S} = \mathbf{E} \times \mathbf{H} \quad (1.62)$$

The total rate of change of stored electromagnetic energy in a closed volume is equal to the integral of the normal component of the Poynting vector over the surface enclosing that volume,

$$P = -\int_{\text{surface}} (\mathbf{E} \times \mathbf{H}) \cdot d\mathbf{s} \quad (1.61)$$

Example 1.5 A coaxial cable, inner conductor radius a, outer conductor of inner radius b, carries a direct current I and has a potential difference between conductors V. By finding expressions for the electric field intensity and the magnetic field intensity in the air space between the conductors, calculate the power flow in the cable from the integral of the Poynting vector across the cross-section of the cable.

Answer Postulate an electrical line charge density of q C/m on the inner conductor of the cable. Apply Gauss's law, eqn. (1.26), to the air space between the conductors. By symmetry, the electric field intensity is radially outwards. Then at any radius r between the two conductors,

$$E = \frac{q}{2\pi r \varepsilon_0}$$

The potential difference between the conductors is found by applying the integral in eqn. (1.9) between the conductors, therefore

$$V = -\int_a^b \mathbf{E} \cdot d\mathbf{l} = -\frac{q}{2\pi\varepsilon_0}\int_a^b \frac{dr}{r} = -\frac{q}{2\pi\varepsilon_0}\ln\frac{b}{a}.$$

Substituting for q,

$$E = \frac{V}{\left(\ln\frac{b}{a}\right)r}$$

Applying Ampère's law, eqn. (1.11), to a circular path enclosing all the current in the centre conductor gives

$$H = \frac{I}{2\pi r}$$

Then the Poynting vector in the air space between the conductors is given by

$$S = EH = \frac{VI}{2\pi r^2\left(\ln\frac{b}{a}\right)}$$

because the E and H fields are orthogonal to one another. The power flow is given by the integral of the Poynting vector across the cross-section of the air space between the conductors. Both E and H lie in the plane of the cross-section of the cable, so that the Poynting vector is directed along the length of the conductors. Therefore,

$$\text{Power} = \int_{\text{area}} \mathbf{S} \cdot d\mathbf{s} = \int_{\text{area}} EH\, ds = \int_a^b \int_0^{2\pi} \frac{VIr\, d\theta\, dr}{2\pi r^2\left(\ln\frac{b}{a}\right)}$$

$$\text{Power} = \frac{2\pi VI}{2\pi\left(\ln\dfrac{b}{a}\right)} \int_a^b \frac{dr}{r} = VI$$

which is the same as the d.c. power carried by the cable, given by the product of the current with the voltage.

1.6 SOURCES AND POTENTIALS

Electrostatics is the study of the forces acting on charges. It becomes convenient to replace the study of Coulomb's law and the inverse square law with the concept of the electric field intensity due to one source acting on the other charge. In the case of a limited number of charges interacting together, there may be little advantage in using the field concept. However, if there is a large number of charges, it is probably much easier to determine the field intensity due to the charges and then to use that field intensity to determine the force acting on yet another charge. So far in this chapter, in the development of Maxwell's equations little has been said about the sources of the fields. The electric field intensity may be determined by the application of Gauss's law and the magnetic field intensity by the application of Ampère's law. In their integral form and for a field system having a certain symmetry, they can be used to determine the field due to charges or currents at a distance. In their differential form, however, Maxwell's equations only relate the field quantities to the charges and currents at the same point. It is necessary to use some additional relationships to calculate the field quantities in terms of the charges and currents some distance away.

Consider the electrostatic field. Gauss's law gives the relationship between the electric flux density and a system of electrostatic charges. Then the electric field intensity is given in terms of the electric potential difference according to eqn. (1.23). Substituting from that equation into Gauss's law, eqn. (1.25) gives the electrostatic potential in terms of the charges in eqn. (1.29):

$$\nabla^2 V = -\frac{\rho}{\varepsilon_0} \tag{1.29}$$

For a single point charge in isolation, the potential is given by

$$V = \frac{Q}{4\pi\varepsilon_0 r} \tag{1.63}$$

Any distributed charge may be considered to be a large number of individual point charges so that the potential field due to the distributed charges is the integral of a large number of expressions similar to eqn. (1.63):

$$V = \frac{1}{4\pi\varepsilon_0} \int_{\text{volume}} \frac{\rho}{r} \, dv' \qquad (1.64)$$

where the volume for the integration is the volume containing the charges and not the volume containing the point at which V is measured. Therefore the volume of integration has been primed. Equation (1.64) is now a solution to Poisson's equation, eqn. (1.29).

The voltage is a scalar potential, that is, it is a unique quantity for each point in space similar to the height on a map. By the principle of superposition, the voltage is also a quantity which is the sum of the voltages due to each individual separate charge. By the application of Coulomb's law, the effect of a number of individual charges involves the vector summation of a number of forces. However, the voltages due to each individual charge may be added together simply to give the voltage due to all the charges working together. Potential is an important concept in many aspects of field theory and a scalar potential such as voltage may be used to determine the field quantities for any conservative field where an equation similar to eqn. (1.10) applies.

Voltage is a scalar potential which may be used to calculate the electric field intensity using eqn. (1.23). The value of the voltage in the electric field is calculated from a knowledge of the electrostatic charges, using eqn. (1.64). Now we need a similar relationship to determine the magnetic field intensity from a knowledge of the electric currents.

Consider the vector potential defined by eqn. (1.46),

$$\mathbf{B} = \text{curl}\,\mathbf{A} \qquad (1.46)$$

where $\mathbf{A}$ is the magnetic vector potential. It does not have the advantage of the scalar potential, V, that a purely scalar addition of values at each point gives the effect of a number of different sources, but it will be seen to have a number of other similarities to V which causes $\mathbf{A}$ to be called a potential. Substitution from eqn. (1.46) into Ampère's law, eqn. (1.40), gives

$$\text{curl}\,\text{curl}\,\mathbf{A} = \mu_0 \mathbf{J} \qquad (1.65)$$

The expansion of curl curl is given in eqn. (1.46); therefore

$$\text{grad}\,\text{div}\,\mathbf{A} - \nabla^2 \mathbf{A} = \mu_0 \mathbf{J} \qquad (1.66)$$

The definition of $\mathbf{A}$ given in eqn. (1.46) is incomplete in that its curl has been defined but that does not define its divergence. Therefore, in a static system we define

$$\text{div}\,\mathbf{A} = 0 \qquad (1.67)$$

Then eqn. (1.66) becomes Poisson's equation:

$$\nabla^2 \mathbf{A} = -\mu_0 \mathbf{J} \qquad (1.68)$$

Although eqn. (1.68) is a vector equation, it is only a shorthand for three similar equations in a rectangular system of coordinates,

$$\nabla^2 A_x = -\mu_0 J_x \tag{1.69}$$

and similar equations in the y- and z-directed components of **A** and **J**. Equation (1.69) is similar to eqn. (1.29) and has a solution similar to eqn. (1.64),

$$A_x = \frac{\mu_0}{4\pi} \int_{\text{volume}} \frac{J_x}{r} \, dv' \tag{1.70}$$

and there are similar expressions for the y- and z-directed components of **A** and **J**. These may be combined into one vector equation, which is a solution to eqn. (1.68),

$$\mathbf{A} = \frac{\mu_0}{4\pi} \int_{\text{volume}} \frac{\mathbf{J}}{r} \, dv' \tag{1.71}$$

where the volume for the integration is the volume containing the electric currents and not the volume containing the point at which **A** is measured. Therefore the volume of integration has been primed. The vector potential is a vector which is parallel to the electric current generating the magnetic field. Poisson's equation and its solution which occur in the calculation of both V and **A** are the reason why **A** is called a potential. The electromagnetic scalar and vector potentials are useful in calculating the electric and magnetic field quantities from a knowledge of the sources of the field.

Elementary electrostatic theory uses Gauss's law to calculate the electric flux density in electrostatic field systems having symmetry. However, very few practical systems are conveniently symmetrical and it is necessary first to calculate the potential values in the field as a route to determining the flux density. Similarly, elementary magnetic field theory uses Ampère's circuital law to calculate the magnetic field intensity in magnetic field systems having symmetry. In systems without the convenient symmetry, it is necessary first to calculate the vector potential values in the field as the route to determining the flux density and the field intensity.

SUMMARY

V is the electrostatic scalar potential:

$$\nabla^2 V = -\frac{\rho}{\varepsilon_0} \tag{1.29}$$

$$V = \frac{1}{4\pi\varepsilon_0} \int_{\text{volume}} \frac{\rho}{r} \, dv' \tag{1.64}$$

$$\mathbf{E} = -\text{grad } V \tag{1.23}$$

A is the magnetic vector potential:

$$\nabla^2 \mathbf{A} = -\mu_0 \mathbf{J} \tag{1.68}$$

$$\mathbf{A} = \frac{\mu_0}{4\pi} \int_{\text{volume}} \frac{\mathbf{J}}{r} \, dv' \tag{1.71}$$

$$\mathbf{B} = \text{curl } \mathbf{A} \tag{1.46}$$

Example 1.6 Find an expression for the electric field intensity on the axis of a circular disk of radius a that has a uniform surface charge density σ.

Answer The geometry of the problem is shown in Figure 1.6. Because there is no overall symmetry, it is not possible to apply Gauss's law to the solution to this problem. It is necessary to calculate the electrostatic potential as the means to finding the field intensity. It is given by eqn. (1.64). Consider an elemental circle of charge in the plane of the disk as shown in Figure 1.6. Then the distance r in eqn. (1.64) is given by $r = \sqrt{(z^2 + R^2)}$ and the total charge in the circle is $\sigma 2\pi R \, dR$. Then to complete eqn. (1.64) it is only necessary to integrate along the radius of the disk:

$$V = \frac{1}{4\pi\varepsilon_0} \int_0^a \frac{\sigma 2\pi R \, dR}{\sqrt{(z^2 + R^2)}} = \frac{\sigma}{2\varepsilon_0} [\sqrt{(z^2 + a^2)} - |z|]$$

$$\mathbf{E} = -\text{grad } V = -\mathbf{U}_z \frac{\partial V}{\partial z}$$

$$= \mathbf{U}_z \frac{\sigma}{2\varepsilon_0} \left[1 - \frac{z}{\sqrt{(z^2 + a^2)}}\right], \quad z > 0$$

$$= \mathbf{U}_z \frac{\sigma}{2\varepsilon_0} \left[-1 - \frac{z}{\sqrt{(z^2 + a^2)}}\right], \quad z < 0$$

The determination of the electric field intensity away from the axis of the disk is a much more difficult problem because there is no longer any symmetry aiding the solution of the integral.

Example 1.7 An electric current I flows in a wire in the form of a square of side $2a$. Find an expression for the magnetic field intensity at the centre of the square.

Answer We shall consider each arm of the square independently. Then, by symmetry, the total field intensity is four times that due to one side of the square. The field intensity is found by first calculating the value of the vector potential. The geometry for one wire of length $2a$ is shown in Figure 1.7. Then the vector potential is found from eqn. (1.71),

$$\mathbf{A} = \frac{\mu_0 I}{4\pi} \int_{-a}^{a} \frac{dx}{\sqrt{(x^2 + R^2)}} = \frac{\mu_0 I}{4\pi} |\ln[x + \sqrt{(x^2 + R^2)}]|_{-a}^{a}$$

$$= \frac{\mu_0 I}{4\pi} \ln\left[\frac{\sqrt{(a^2 + R^2)} + a}{\sqrt{(a^2 + R^2)} - a}\right]$$

SOURCES AND POTENTIALS 33

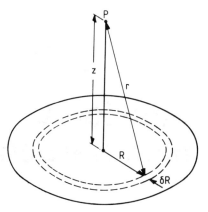

Figure 1.6 The geometry of Example 1.6, a point on the axis of a circular disk of radius a.

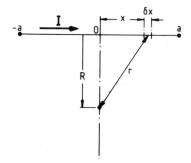

Figure 1.7 Part of the answer to Example 1.7, showing the field at a radial distance from the centre of a wire of finite length.

The **A** vector is parallel to the current I. Curl gives rise to components perpendicular to the vector being curled. However, because the field is symmetrical about the axis of the wire, the only component of **B** is circumferential about the wire. The radial component is zero. Therefore, from eqn. (1.46),

$$\mathbf{B} = -\mathbf{U}_\theta \frac{\partial}{\partial R}\left(\frac{\mu_0 I}{4\pi} \ln\left[\frac{\sqrt{(a^2 + R^2)} + a}{\sqrt{(a^2 + R^2)} - a}\right]\right)$$

$$= \mathbf{U}_\theta \frac{\mu_0 I a}{2\pi R\sqrt{(a^2 + R^2)}} = \mathbf{U}_\theta \frac{\mu_0 I}{2\pi\sqrt{2}\, a}$$

when $R = a$ at the centre of the square. It is not possible to substitute $R = a$ earlier because of the need to differentiate the expression for **A** in order to find **B**. The contribution to the total field from each side of the square is equal; therefore the field intensity at the centre of the square is given by

$$\mathbf{H} = \mathbf{U}_\theta \frac{I\sqrt{2}}{\pi a}$$

CHAPTER 2

Transmission Lines I: Switching Surges

Aims: The aim of this chapter is to introduce the reader to the effects of time delay and switching surges in transmission lines.

2.1 INTRODUCTION

A transmission line consists of any system of conductors that can be used to transmit electrical energy between two or more points. When a voltage source is connected to the input of a transmission line, the potential difference all along the line cannot rise instantaneously to that of the source. Time is needed for the transfer of energy corresponding to the potential difference between the conductors of the line. The special theory of relativity also indicates that an instantaneous change of potential along the whole length of the line is impossible. No electrical signal can be transmitted at a speed greater than the speed of light and time is needed for charge to travel along a transmission line.

A circuit connected by a transmission line of length l is shown in Figure 2.1. Two ammeters are connected in the circuit, one at the source and one at the load. On closing the switch, a current will flow in the ammeter A_1 immediately but the current will only flow in A_2 some time later. The first flicker in the ammeter A_2 will occur at a time l/c after the switch was closed, where c is the speed of light. The steady-state conditions for the current flowing in A_2 occur sometime later than that.

Consider any two-wire transmission line of infinite length as shown in

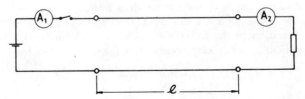

Figure 2.1 A source and load connected by a transmission line of length l.

INTRODUCTION   35

Figure 2.2. The only condition that needs to be specified is that the line maintains a constant cross-section throughout its length. If it is connected to a voltage source as shown in Figure 2.2, a current will flow into the line in order to charge up the capacitance between the wires. The current will continue to flow for as long as the voltage source is connected because the line is of infinite length and can never be completely charged. The current will always flow even if there is no leakage conductance between the wires of the line. The ratio

$$\frac{V}{I} = Z_0 \tag{2.1}$$

is a constant for any particular line and is called the *characteristic impedance* of the line. The input impedance of any infinitely long line is its characteristic impedance. If a short length of line is cut off the input of an infinitely long line, the input impedance of the remaining infinitely long line will be the same characteristic impedance. Similarly, if a short length of line is terminated in an infinite length of the same line, it will behave as an infinite line and its input impedance will be the characteristic impedance. The infinite line may be replaced by its characteristic impedance without disturbing the electrical conditions on the short line, so that a short line terminated in its characteristic impedance behaves like an infinite line. The equivalence is shown in Figure 2.3.

It takes time for electrical charge to travel along a transmission line. For an infinite line, there can be no reflections from the end of the line because the signal never reaches the end of the line. Similarly there can be no reflections from the end of the line that is terminated in its characteristic impedance because such a line behaves as if it were a line of infinite length.

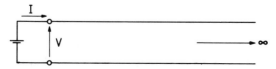

Figure 2.2 A current flows into a transmission line of infinite length (see p. ix).

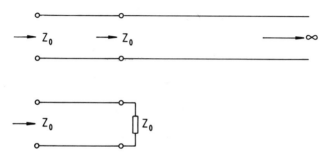

Figure 2.3 Showing the equivalence between an infinite line and a short line terminated in its characteristic impedance. (see p. ix).

SUMMARY

The input impedance of an infinitely long line of constant cross-section is constant and is called the *characteristic impedance* of the line.

A short length of line terminated in its characteristic impedance behaves like an infinite line.

There are no reflections from the end of a line terminated in its characteristic impedance.

2.2 TRANSMISSION LINE EQUATIONS

Consider any two-wire transmission line. The wire of the line has a series resistance, R Ω/m, and a series inductance, L H/m. There is also a shunt capacitance, C F/m, and a shunt conductance, G S/m, between the two wires of the line. In many practical lines, the shunt conductance is negligible but it is included here for completeness. The effect of a short length of line, δz, is shown in Figure 2.4. The current and voltage on the line are also shown in Figure 2.4. Although it is obvious from the circuit diagram that the current and voltage decrease as z increases along the line, both are shown to increase with increase of z so that the signs are correct in the mathematics.

Applying Kirchhoff's current law to the circuit shown in Figure 2.4 gives

$$I - (I + \delta I) = G\,\delta z\,V + C\,\delta z\,\frac{\partial V}{\partial t}$$

In the limit, as $\delta z \to 0$, this equation reduces to

$$-\frac{\partial I}{\partial z} = GV + C\frac{\partial V}{\partial t} \qquad (2.2)$$

Applying Kirchhoff's law to the voltage drop across R and L and neglecting the second-order small terms in $\delta I\,\delta z$ gives

$$V - (V + \delta V) = R\,\delta z\,I + L\,\delta z\,\frac{\partial I}{\partial t}$$

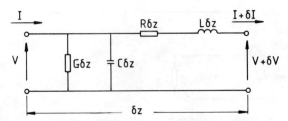

Figure 2.4 Equivalent circuit of a short length, δ_z, of a transmission line.

TRANSMISSION LINE EQUATIONS

Similarly, this equation reduces to

$$-\frac{\partial V}{\partial z} = RI + L\frac{\partial I}{\partial t} \quad [\text{divide} =] \quad (2.3)$$

Since most transmission lines are low loss and the series resistance, R, and the shunt conductance, G, are very small, the assumption will be made that

$$R = 0 \quad \text{and} \quad G = 0$$

so simplifying eqns. (2.2) and (2.3). They become

$$\frac{\partial I}{\partial z} = -C\frac{\partial V}{\partial t} \quad (2.4)$$

and

$$\frac{\partial V}{\partial z} = -L\frac{\partial I}{\partial t} \quad (2.5)$$

The current I will be eliminated from these two equations. First differentiate eqn. (2.5) with respect to z,

$$\frac{\partial^2 V}{\partial z^2} = -L\frac{\partial^2 I}{\partial t \partial x}$$

and differentiate eqn. (2.4) with respect to t,

$$\frac{\partial^2 I}{\partial t \partial z} = -C\frac{\partial^2 V}{\partial t^2}$$

Substituting one into the other gives

$$\frac{\partial^2 V}{\partial z^2} = LC\frac{\partial^2 V}{\partial t^2} \quad (2.6)$$

It is left to the reader to eliminate the voltage V from eqns. (2.4) and (2.5) by a similar process to show that

$$\frac{\partial^2 I}{\partial z^2} = LC\frac{\partial^2 I}{\partial t^2} \quad (2.7)$$

Equations (2.6) and (2.7) are wave equations and a possible solution has the form

$$\boxed{V = f(z \pm vt)} \quad (2.8)$$

f is any arbitrary function which is usually a sinusoidal function or a combination of exponential functions. Differentiating eqn. (2.8) gives

$$\frac{\partial V}{\partial z} = f'(z \pm vt)$$

$$\frac{\partial^2 V}{\partial z^2} = f''(z \pm vt)$$

$$\frac{\partial V}{\partial t} = \pm vf'(z \pm vt)$$

$$\frac{\partial V}{\partial t^2} = v^2 f''(z \pm vt)$$

where the prime denotes differentiation with respect to the argument. Therefore

$$v^2 \frac{\partial^2 V}{\partial z^2} = \frac{\partial^2 V}{\partial t^2} \tag{2.9}$$

By comparison with eqn. (2.6)

$$v^2 = \frac{1}{LC} \tag{2.10}$$

and v is the velocity of the wave.

Equation (2.6) is a second-order partial differential equation whose solution contains two arbitrary functions. Therefore the most general solution of eqn. (2.6) is

$$V = f(z + vt) + g(z - vt) \tag{2.11}$$

where f and g are arbitrary functions whose shape is determined by the boundary conditions in z and t. Consider just one of these solutions,

$$V = g(z - vt)$$

As shown by Figure 2.5, $g(z - vt)$ is a wave of fixed shape travelling in the positive z direction with a velocity v. Similarly, $f(z + vt)$ is a wave of fixed shape travelling in the negative z direction with a velocity v. The total voltage on the line is given by the sum of these two waves.

The electric current flowing in the line is derived by substituting eqn. (2.11) into eqn. (2.4). Therefore

$$\frac{\partial I}{\partial z} = -C \frac{\partial V}{\partial t} = -C[vf'(z + vt) - vg'(z - vt)]$$

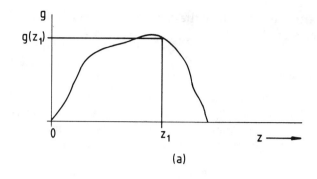

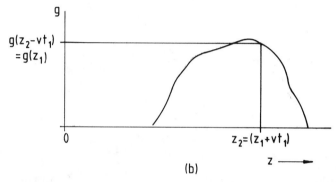

Figure 2.5 An arbitrary wave $g(z)$ travelling in the positive z-direction with a velocity v: (a) time $t = 0$, (b) time $t = t_1$.

Integrating with respect to z gives

$$I = Cv[-f(z + vt) + g(z - vt)]$$

and substituting from eqn. (2.10) gives

$$I = \sqrt{\frac{C}{L}}[-f(z + vt) + g(z - vt)] \quad (2.12)$$

The constant $\sqrt{(L/C)}$ has the dimensions of impedance and is called the *characteristic impedance* of the line. It gives a numerical ratio between the current wave and the voltage wave on the line and is the same as the characteristic impedance of the infinitely long line discussed in Section 2.1. Therefore

$$Z_0 = \sqrt{\frac{L}{C}} \quad (2.13)$$

For the forward wave, $V = g(z - vt)$,

$$V = Z_0 I$$

40 TRANSMISSION LINES I: SWITCHING SURGES

and the current wave is exactly the same shape as the voltage wave. For the reverse wave, $V = f(z + vt)$ and

$$V = -Z_0 I$$

Again the current due to the reverse wave is a wave of exactly the same shape as the voltage wave but it is now flowing in the negative z direction, hence giving rise to the negative sign in the above equation. For both the forward and reverse waves, the current flows in the direction of travel of the wave.

SUMMARY

The transmission line equations are

$$-\frac{\partial I}{\partial z} = GV + C\frac{\partial V}{\partial t} \tag{2.2}$$

$$-\frac{\partial V}{\partial z} = RI + L\frac{\partial I}{\partial t} \tag{2.3}$$

For the low-loss line, we consider that $R = 0$ and $G = 0$ and eqns. (2.2) and (2.3) can be solved to give the wave equations

$$\frac{\partial^2 V}{\partial z^2} = LC\frac{\partial^2 V}{\partial t^2} \tag{2.6}$$

$$\frac{\partial^2 I}{\partial z^2} = LC\frac{\partial^2 I}{\partial t^2} \tag{2.7}$$

The solution of eqn. (2.6) is

$$V = f(z + vt) + g(f - vt) \tag{2.11}$$

where

$$v^2 = \frac{1}{LC} \tag{2.10}$$

and the related solution for the current is given by

$$Z_0 I = -f(z + vt) + g(z - vt)$$

where

$$Z_0 = \sqrt{\frac{L}{C}} \tag{2.13}$$

Example 2.1 The line constants of an open-wire telephone line are given by $R = 6.3 \times 10^{-3}$ Ω/m, $L = 2.3 \times 10^{-6}$ H/m, $C = 4.9 \times 10^{-12}$ F/m and $G = 0.2 \times 10^{-9}$ S/m. Calculate its characteristic impedance and wave velocity.

Answer The velocity is given by eqn. (2.10). Therefore

$$v^2 = \frac{1}{LC} = \frac{1}{2.3 \times 10^{-6} \times 4.9 \times 10^{-12}} = 8.87 \times 10^{16}$$

Therefore

$$v = 2.98 \times 10^8 \text{ m/s}$$

which is approximately equal to the speed of light.
The characteristic impedance is given by eqn. (2.13). Therefore

$$Z_0 = \sqrt{\frac{L}{C}} = \sqrt{\frac{2.3 \times 10^{-6}}{4.9 \times 10^{-12}}} = 685 \text{ Ω}$$

PROBLEM

2.1 A coaxial cable has the following properties: $R = 3.1 \times 10^{-3}$ Ω/m, $L = 0.24 \times 10^{-6}$ H/m, $G = 3.8 \times 10^{-6}$ S/m and $C = 94 \times 10^{-12}$ F/m. Find its characteristic impedance and wave velocity.

[50.5 Ω, 2.1×10^8 m/s]

2.3 LINE TERMINATION

So far we have considered only an infinite line and a line terminated in its characteristic impedance. Any other termination causes a signal to be reflected from the junction or the termination, which gives rise to backward waves on the line. Consider a junction between two different lines as shown in Figure 2.6. This is a transmission line of characteristic impedance Z_0 terminated by another line of characteristic impedance Z_t. If we take the

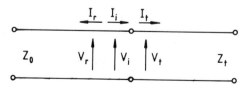

Figure 2.6 A line of characteristic impedance Z_0 connected to a line of characteristic impedance Z_t.

TRANSMISSION LINES I: SWITCHING SURGES

forward current flow as positive, the electrical signals at the junction are

an incident wave, V_i and I_i where $V_i = Z_0 I_i$

which gives rise to

a transmitted wave V_t and I_t where $V_t = Z_t I_t$

and

a reflected wave V_r and I_r where $V_r = -Z_0 I_r$

and the negative sign makes allowance for the fact that the current of the reflected wave is flowing in the direction of travel of that wave.

At the junction, the total current flowing into the junction is zero and the total voltage on each side of the junction is equal. Therefore

$$I_i + I_r = I_t \tag{2.14}$$

and

$$V_i + V_r = V_t \tag{2.15}$$

Substituting the voltages into eqn. (2.14) gives

$$\frac{V_i}{Z_0} - \frac{V_r}{Z_0} = \frac{V_t}{Z_t}$$

which reduces to

$$V_i - V_r = \frac{Z_0}{Z_t} V_t \tag{2.16}$$

Equations (2.15) and (2.16) make a pair of simultaneous equations which may be solved for V_i and V_r, giving

$$2V_i = V_t \left(1 + \frac{Z_0}{Z_t}\right) \tag{2.17}$$

$$2V_r = V_t \left(1 - \frac{Z_0}{Z_t}\right) \tag{2.18}$$

Taking the ratio of eqns. (2.18) and (2.17) gives

$$\frac{V_r}{V_i} = \frac{Z_t - Z_0}{Z_t + Z_0} \tag{2.19}$$

The ratio in eqn. (2.19) is called the *reflection coefficient* and is denoted by the symbol ρ,

$$\rho = \frac{\text{reflected voltage}}{\text{incident voltage}} = \frac{V_r}{V_i} = \frac{Z_t - Z_0}{Z_t + Z_0} \qquad (2.20)$$

The current reflection coefficient is defined similarly and is obtained by substitution into eqn. (2.19),

$$\rho_c = \frac{I_r}{I_i} = -\rho = \frac{Z_0 - Z_t}{Z_0 + Z_t} \qquad (2.21)$$

The negative sign occurs because the current in the reflected wave is flowing in the opposite direction to the current in the incident wave.

From eqn. (2.17) we obtain

$$V_t = \left(\frac{2Z_t}{Z_t + Z_0}\right) V_i \qquad (2.22)$$

Therefore

$$I_t = \left(\frac{2}{Z_t + Z_0}\right) V_i$$

and

$$I_t = \left(\frac{2Z_0}{Z_t + Z_0}\right) I_i \qquad (2.23)$$

The ratio given in eqn. (2.22) is defined similarly to the reflection coefficient to be the *transmission coefficient*,

$$\mathcal{T} = \frac{\text{transmitted voltage}}{\text{incident voltage}} = \frac{V_t}{V_i} = \frac{2Z_t}{Z_0 + Z_t} \qquad (2.24)$$

It has already been shown in Section 2.1 that a line may be replaced by its characteristic impedance. Therefore the terminating line in Figure 2.6 may be replaced by its characteristic impedance as shown in Figure 2.7. Then we have the situation of a line of characteristic impedance Z_0 terminated by an

Figure 2.7 A line of characteristic impedance Z_0 terminated with an impedance Z_t.

impedance Z_t. As far as the incident wave is concerned, it does not know whether the line is terminated by another line or by an impedance of the same value. So eqns. (2.20) and (2.21) also apply to any line terminated in an impedance other than its characteristic impedance. The transmitted portion of the wave is absorbed in the terminating impedance.

SUMMARY

For a line connected to another line or a line terminated in an impedance, a reflected wave is generated proportional to the incident wave.

$$\text{Reflection coefficient} = \rho = \frac{Z_t - Z_0}{Z_t + Z_0} \quad (2.20)$$

$$\text{Current reflection coefficient} = \rho_c = -\rho = \frac{Z_0 - Z_t}{Z_0 + Z_t} \quad (2.21)$$

$$\text{Transmission coefficient} = \mathcal{T} = \frac{2Z_t}{Z_t + Z_0} \quad (2.24)$$

The transmitted portion of the wave is given by

$$V_t = \mathcal{T} V_i$$

$$I_t = \left(\frac{2Z_0}{Z_t + Z_0}\right) I_i \quad (2.23)$$

and the transmitted power is absorbed in the terminating impedance.

Example 2.2 A 50 Ω transmission line is connected to a 75 Ω transmission line. Find the voltage and current reflection coefficients and the transmission coefficient due to the junction.

Answer The solution is given by substitution into eqns. (2.20)–(2.24) using the values, $Z_0 = 50\,\Omega$, $Z_t = 75\,\Omega$:

$$\rho = \frac{Z_t - Z_0}{Z_t + Z_0} = \frac{25}{125} = 0.20$$

$$\rho_c = \frac{Z_0 - Z_t}{Z_0 + Z_t} = \frac{-25}{125} = -0.20$$

$$\mathcal{T} = \frac{2Z_t}{Z_t + Z_0} = \frac{150}{125} = 1.20$$

Example 2.3 A measuring instrument has an input impedance of 60 Ω resistive. A coaxial cable of characteristic impedance 75 Ω is used to make connection to the measuring instrument. Find the reflection coefficient at the junction between the cable and the measuring instrument.

Answer The measuring instrument appears as a terminating impedance of 60 Ω on the end of the coaxial cable, so that the reflection coefficient formula in eqn. (2.20) is still valid:

$$\rho = \frac{Z_t - Z_0}{Z_t + Z_0} = \frac{60 - 75}{60 + 75} = -0.11$$

PROBLEMS

2.2 A transmission line of characteristic impedance 50 Ω is terminated in a resistor of 100 Ω. Find the reflection coefficient at the termination.

[0.33]

2.3 An open-wire telephone line has the following measured parameters: $R = 6.5$ Ω/km, $L = 2.3$ mH/km, $C = 0.0052$ μF/km, $G = 0.50$ μS/km. Find the characteristic impedance and wave velocity of the line. The line is connected to a receiving instrument which presents an impedance of 600 Ω to the line. Find the reflection coefficient at the receiving end of the line.

[665 Ω, 2.9×10^8 m/s; 0.05]

2.4 VOLTAGE STEP

When a steady voltage is switched onto a line, that voltage does not appear everywhere on that line instantaneously. A voltage surge travels along the line at the wave velocity of the line. At the same time, a steady current flows into the line in order to charge up the capacitance of the line. Consider a line connected through its characteristic impedance to a battery as shown in Figure 2.8 with its opposite end open circuited. When the switch is closed, the battery sees the characteristic impedance of the line in series with the source resistance R_s. If $R_s = Z_0$, $\frac{1}{2}V$ appears as the voltage on the line and a voltage and current surge travels along the line at the wave velocity of the line as shown by Figure 2.8(b). The voltage and current surges travel together and before they arrive the line does not know that the switch has been closed. When the surge reaches the end of the line, a reflected surge is generated which is governed by the reflection coefficient and transmission coefficient at the end of the line. For an open circuit, $Z_t = \infty$. Therefore

$$\rho = 1, \quad V_r = V_i, \quad I_r = -I_i, \quad V_t = 2V_i, \quad I_t = 0$$

where the notation is taken from Figure 2.6. The reflected voltage and

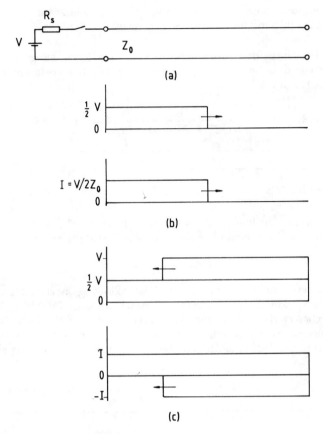

Figure 2.8 Voltage and current surge on an open-circuited line: (a) circuit; (b) before the surge reaches the end of the line; (c) the reflected surge before it reaches the source.

current surge is shown in Figure 2.8(c). It is seen that the reflected voltage adds to the voltage already on the line to give a voltage of V at the termination. The reflected current surge is negative, so that there is no current flowing into the termination. The final situation at the termination is that of a simple d.c. connection to an open circuit.

The situation shown in Figure 2.8(c) is at some time later than the situation shown in Figure 2.8(b). It takes a small, but finite, time for the switching surge to travel along the line and for the reflected surge to travel back. All that time, current is flowing into the line from the battery to charge up the capacitance of the line. During the forward travelling voltage step shown in Figure 2.8(b) it is easy to see that the current flows at a rate given by the product of voltage, capacitance per meter and velocity. During the passage of the reflected voltage step shown in Figure 2.8(c), charge is being accumulated at the same rate and the same current continues to flow until that voltage step has passed. At any point on the line, the voltage and the current remain constant until the voltage step arrives. When the surge returns to the source, the source appears to have the characteristic impedance of the line and no further reflections occur. The switching surge is over and the static d.c. conditions prevail on the line.

The similar conditions for a line terminated in a short circuit are shown in Figure 2.9. The effect of the initial switching surge as shown in Figure 2.9(b) is the same as that shown in Figure 2.8(b) for the open-circuited line. However, the terminating conditions are different. They are $Z_t = 0$. Therefore

$$\rho = -1, \quad V_r = -V_i, \quad I_r = I_i, \quad V_t = 0, \quad I_t = 2I_i$$

where the notation is taken from Figure 2.6. For this situation, the reflected voltage and current surge shown in Figure 2.9(c) is the dual of that for the open-circuited line as shown in Figure 2.8(c). The final situation is that of a simple d.c. connection to a short circuit.

If the line is terminated in its characteristic impedance, there is no reflected surge, and the final current and voltage on the line are the same as those shown in Figure 2.8(b) and Figure 2.9(b) even after the voltage step has reached the end of the line.

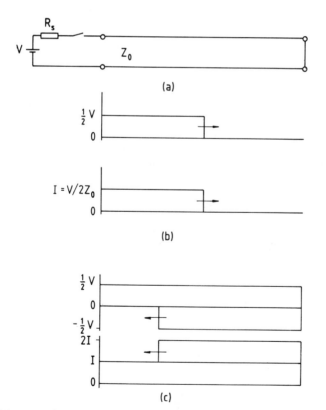

Figure 2.9 Voltage and current surge on a short-circuited line: (a) circuit; (b) before the surge reaches the end of the line; (c) the reflected surge before it reaches the source.

In the circuit described by Figures 2.8 and 2.9, the source impedance was assumed to be the same as the characteristic impedance of the line. If it is not, then the voltage is divided between the source impedance and the line in proportion to their impedances. Consider the circuit shown in Figure 2.10(a). As soon as the switch is closed, the circuit behaves as the circuit shown in Figure 2.10(b). The voltage V is developed across the line which is given by

$$V = \frac{Z_0}{Z_0 + R_s} V_0 \qquad (2.25)$$

If the line is not infinitely long or terminated in its characteristic impedance, any reflected voltage step sees the line as terminated in the source impedance, R_s. As far as the source termination is concerned, the voltage source is replaced by its Thévenin equivalent impedance, and there will be a source reflection coefficient given by

$$\rho = \frac{R_s - Z_0}{R_s + Z_0} \qquad (2.26)$$

If a line is connected to two other lines in a junction as shown in Figure 2.11(a), the junction appears as a terminating impedance given by the characteristic impedances of the two lines in parallel,

$$Z_j = \frac{Z_{02} Z_{03}}{Z_{02} + Z_{03}} \qquad (2.27)$$

The reflection coefficient is given by

$$\rho = \frac{Z_j - Z_{01}}{Z_j + Z_{01}} \qquad (2.28)$$

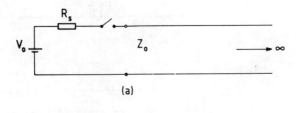

(a)

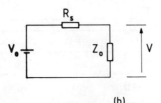

(b)

Figure 2.10 The effect of the source impedance on the initial voltage on the line: (a) circuit; (b) equivalent circuit at the instant after the switch is closed.

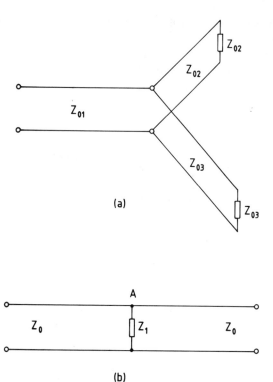

Figure 2.11 (a) A junction between three transmission lines; (b) an impedance in shunt across a transmission line.

The transmitted voltage appears across Z_{02} and Z_{03} in parallel. If an impedance is added in shunt across a line as shown in Figure 2.11(b), the calculation is the same as for the junction between three lines. The effective impedance at the point A is given by

$$Z = \frac{Z_1 Z_0}{Z_1 + Z_0}$$

and the reflection coefficient is given by

$$\rho = \frac{Z - Z_0}{Z + Z_0} = \frac{-Z_0}{2Z_1 + Z_0} \qquad (2.29)$$

If any switching surge is reflected from both ends of the line, the situation is much more complicated than the situation described in Figures 2.8 and 2.9. It is best described by a *space–time diagram*. Distance along the transmission line is plotted horizontally across a page, and time following the initial switch closure is plotted vertically on the same page. The progression of any particular voltage step on the line is plotted as a line on the space–time diagram. The size of each voltage step must be indicated for each line on the diagram. The use of the space–time diagram is best demonstrated by a simple example.

Consider a voltage source of 180 V, internal impedance $0.8Z_0$, connected to a load impedance of $4.0Z_0$ through a line of characteristic impedance Z_0 of length l, as shown in Figure 2.12(a). When the switch is closed, the switching step will travel along the line at the wave velocity of the line, v. The initial voltage step on the line is given by eqn. (2.25),

$$V = \frac{Z_0}{Z_0 + 0.8Z_0} 180 = 100 \text{ V}$$

The time of propagation of the voltage step along the line is $T = l/v$. When the voltage step reaches the end of the line, it is reflected with a reflection

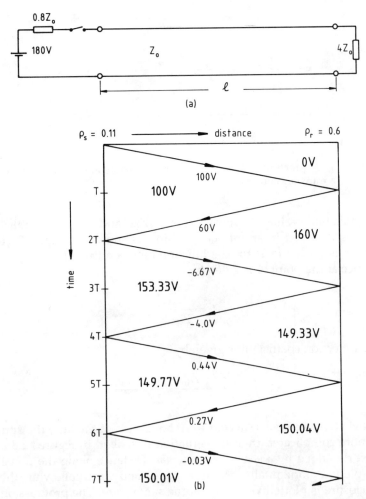

Figure 2.12 Switching surges on a transmission line: (a) circuit diagram; (b) space–time diagram.

coefficient given by

$$\rho_r = \frac{4Z_0 - Z_0}{4Z_0 + Z_0} = \frac{3Z_0}{5Z_0} = 0.6$$

As the reflection coefficient is positive, the reflected voltage step increases the voltage on the line. When the reflected voltage step reaches the source, it is further reflected with a reflection coefficient given by

$$\rho_s = \frac{0.8Z_0 - Z_0}{0.8Z_0 + Z_0} = \frac{-0.2Z_0}{1.8Z_0} = -\frac{1}{9} = -0.111$$

As this time the reflection coefficient is negative, the reflected voltage now decreases the total voltage on the line. As the voltage step is reflected from each end of the line in turn, it gets smaller and the voltage on the line approaches the static value of 150 V. The space–time diagram is shown in Figure 2.12(b). It is seen that the voltage on the line reaches within 1 V of the static value after $4T$ and within 0.1 V of the static value after $6T$. For most circuits the effects of these switching surges may safely be ignored. However, in long lines or on high-speed computers, switching delays may be significant and in high-voltage systems allowance must be made for the highest voltage reached during switching.

SUMMARY

When a voltage source is connected to a transmission line, a voltage step travels along the line at the wave velocity of the line. The initial voltage on the line is given by

$$V = \frac{Z_0}{Z_0 + R_s} V_0 \quad (2.25)$$

The voltage step is reflected from the end of the line unless the line is terminated in its characteristic impedance. The reflected step will see a source reflection coefficient given by

$$\rho_s = \frac{R_s - Z_0}{R_s + Z_0} \quad (2.26)$$

At the junction of three lines, each line sees the characteristic impedance of the other two lines in parallel

$$Z_j = \frac{Z_{02} Z_{03}}{Z_{02} + Z_{03}} \quad (2.27)$$

and the reflection coefficient is given by

$$\rho = \frac{Z_j - Z_{01}}{Z_j + Z_{01}} \quad (2.28)$$

An impedance added in shunt across a line is treated similarly; it generates a reflection coefficient,

$$\rho = \frac{-Z_0}{2Z_1 + Z_0} \quad (2.29)$$

The progress of a switching surge on a transmission line is plotted using a *space–time diagram* where distance along the line is plotted across the page and time is plotted down the page. The size of any voltage step is noted against any line on the diagram and the voltage on the line is noted for any areas on the diagram.

Example 2.4 A source of 160 V, internal impedance $0.6Z_0$, is connected through a transmission line of characteristic impedance, Z_0, length l to a load of impedance $9Z_0$. Given that T is the time taken for the switching surge to make one traverse of the transmission line, plot the voltage at the end of the line and the voltage at the mid-point of the line until time $6T$ after the source is connected to the line.

Answer The circuit is shown in Figure 2.13(a). The initial voltage on the line is given by eqn. (2.25):

$$V = \frac{Z_0}{1.6Z_0} 160 = 100 \text{ V}$$

The source and load relection coefficients are given by

$$\rho_r = \frac{9Z_0 - Z_0}{9Z_0 + Z_0} = 0.8$$

$$\rho_s = \frac{0.6Z_0 - Z_0}{0.6Z_0 + Z_0} = -0.25$$

These results are then plotted on the space–time diagram, as shown in Figure 2.13(b). The required answer is obtained by plotting some of the line voltages taken from the space–time diagram. The answer is given in Figure 2.13(c), where the voltage at the end of the line is given by the solid line and the voltage at the mid-point of the line is given by the dotted line.

Example 2.5 In high-speed digital computers, matched transmission lines are used for interconnections. A clock pulse generator is connected in parallel to two transmission lines of lengths 10 cm and 20 cm. Each line has a characteristic impedance of 75 Ω and a propagation velocity of 2×10^8 m/s and each terminates in a 75 Ω load. The source impedance of 37.5 Ω provides maximum

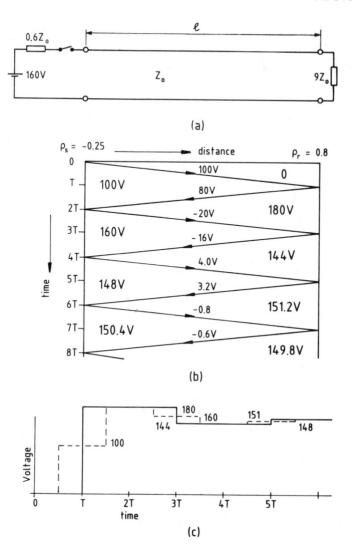

Figure 2.13 Part of the answer to Example 2.4: (a) circuit diagram; (b) space–time diagram; (c) voltage ——— at end of the line, – – – – – at mid-point of the line.

power transfer into the pair of lines. Under normal operating conditions, the generator delivers a pulse of amplitude 5 V and duration 5 ns to each termination. As a result of a fault, the termination of the 20 cm line becomes open circuited. Calculate the signal received at the end of each line for a period of 5 ns after the start of a clock pulse.

Answer The circuit is shown in Figure 2.14(a). At the open circuit, $\rho = 1$. At the beginning, the signal is split equally between the two lines, giving equal voltages of 5 V across 75 Ω. The signal is reflected from the open-circuit end. At the source, the source impedance of 37.5 Ω appears in shunt across the line as

54 TRANSMISSION LINES I: SWITCHING SURGES

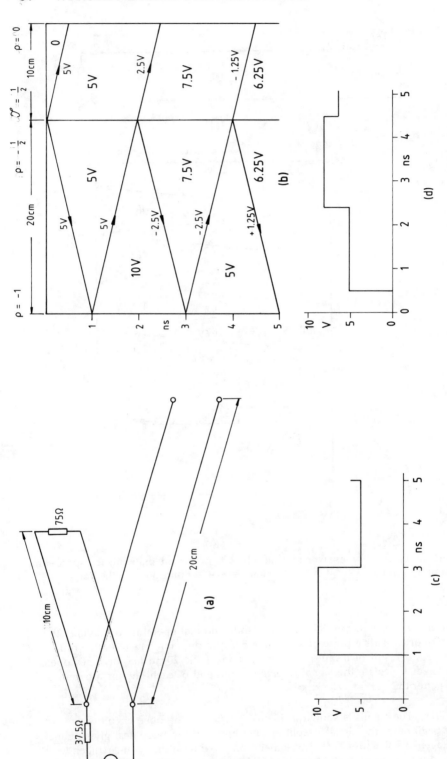

Figure 2.14 Part of the answer to Example 2.5: (a) circuit diagram; (b) space–time diagram; (c) the voltage at the end of the 10 cm line; (d) the voltage at the end of the 20 cm line.

shown in Figure 2.11(b). The effective terminating impedance is given by

$$Z = \frac{37.5 \times 75}{37.5 + 75} = 25 \, \Omega$$

and the reflection coefficient and transmission coefficient at the source are given by

$$\rho = \frac{25 - 75}{25 + 75} = -\frac{1}{2}$$

$$\mathcal{T} = \frac{2 \times 25}{25 + 75} = \frac{1}{2}$$

The time for propagation along 20 cm of the line is given by

$$T = \frac{2 \times 10^{-1}}{2 \times 10^8} = 1 \times 10^{-9} = 1 \text{ ns}$$

The space–time diagram is shown in Figure 2.14(b). The required voltage at the end of each line may be read from the space–time diagram. The voltage at the end of the 20 cm line is given in Figure 2.14(c) and that for the 10 cm line is given in Figure 2.14(d).

Example 2.6 A transcontinental d.c. transmission line 180 km long has the following properties: $R = 70 \times 10^{-6} \, \Omega/\text{m}$, $L = 1.8 \times 10^{-6} \, \text{H/m}$, $G = 3 \times 10^{-12} \, \text{S/m}$, $C = 6 \times 10^{-12} \, \text{F/m}$. It is running light and is connected at one end to a source of negligible impedance and at the other to a load of impedance 1.1 kΩ. Lightning activity causes a voltage surge on the line at a distance of 120 km from the source. This voltage surge has insufficient energy to disturb the circuit breakers at the ends of the line. Given that the voltage surge may be approximated by a flat-topped voltage pulse of duration 30 μs, plot the voltage waveform at the centre point of the line for the first 3 ms after the initiation of the surge.

Answer First calculate the line parameters by substituting into eqns. (2.10) and (2.13)

$$v = \frac{1}{\sqrt{(LC)}} = \frac{1}{\sqrt{(1.8 \times 10^{-6} \times 6 \times 10^{-12})}} = 3 \times 10^8 \text{ m/s}$$

$$Z_0 = \sqrt{\left(\frac{L}{C}\right)} = \sqrt{\left(\frac{1.8 \times 10^{-6}}{6 \times 10^{-12}}\right)} = 548 \, \Omega$$

The velocity is the speed of light in air, or a vacuum, and shows that the transmission line is an open-wire line in air. The time to traverse the line is l/v. Then we have

for 60 km of line $T_1 = 0.2$ ms
for 120 km of line $T_2 = 0.4$ ms
for 180 km of line $T_3 = 0.6$ ms.

The source is an effective short circuit so that the source reflection coefficient is −1. The reflection coefficient at the load is given by

$$\rho_r = \frac{1100 - 548}{1100 + 548} = \frac{1}{3}$$

The circuit diagram and the space–time diagram are given in Figure 2.15(a). The initial pulse divides into two equal-amplitude pulses moving in opposite

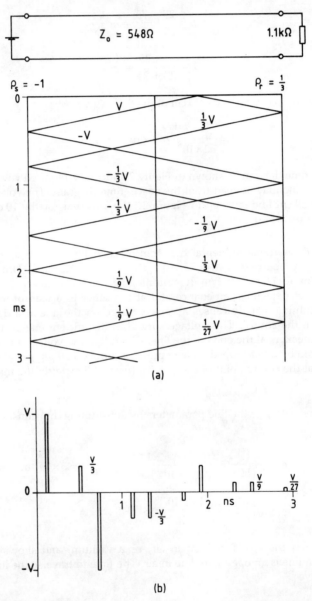

Figure 2.15 Part of the answer to Example 2.6: (a) circuit diagram and space–time diagram; (b) the form of voltage at the mid-point of the line for 3 ms after the voltage pulse is initiated on the line.

directions along the line. All the voltages are represented as fractions of the voltage of the original pulse. The pulse length of 30 μs is short compared with the time for transmission along the line so that each line on the space–time diagram now represents the passage of a 30 μs pulse of the appropriate amplitude. There is no voltage on the line during the intervening intervals. The required answer of the voltage at the centre of the line is given in Figure 2.15(b).

PROBLEMS

2.4 A 56 cm length of cable has a characteristic impedance of 75 Ω and a wave velocity of 2×10^8 m/s. The cable is used to connect a source of internal impedance 40 Ω to an amplifier with an input impedance of 1 MΩ. The source voltage changes from 0 to 5 V at time $t = 0$. Calculate the form of the voltage seen by the amplifier up to time $t = 15$ ns.

2.5 A transmission line, of characteristic impedance Z_0, is terminated in a load equal to $Z_0/2$. A d.c. generator having an e.m.f. of 10 V and an impedance equal to $2Z_0$ is connected to the line. Write out the first four terms in the infinite series for the voltage at a distance of one quarter of the line length from the generator. Express the reflection coefficients numerically.

[1.975 V]

2.6 An overhead transmission line with a characteristic impedance of 400 Ω is connected at one end to two cables, one having a characteristic impedance of 100 Ω and the other 60 Ω. A voltage surge (which may be taken as a voltage step) of 100 kV travels along the overhead line to the junction. Find the magnitude of the voltage and current waves reflected from and transmitted beyond the junction.

[82.8, 17.2 kV; 207, 172, 286 A]

CHAPTER 3

Transmission Lines II: A.C. Effects

Aims: to investigate the effects of phase delay and standing waves on transmission lines and to introduce matching techniques.

3.1 A.C. STEADY STATE

When an a.c. signal is switched onto a transmission line, there is a switching surge on the line similar to that described in Section 2.4 which is dependent on the time in the alternating cycle at which switching occurs. The effect can be very complicated and will not be considered further. However, there is also a phase delay for any a.c. signals on the line or reflected onto the line, due to the time delay of propagation along the line.

Any alternating voltage may be represented mathematically by

$$V = V_m \cos(\omega t + \phi)$$

where V_m is the peak amplitude of the wave and ϕ is the phase delay from some arbitrary datum. The arbitrary datum is usually chosen such that $\phi = 0$. Mathematically, the cosinusoidal dependence of voltage may also be represented as the real part of $\exp[j(\omega t + \phi)]$. However, the statement *the real part of* is usually assumed and the alternating voltage is given by

$$V = V_m \exp[j(\omega t)]$$

and similarly

$$I = I_m \exp[j(\omega t + \phi)]$$

Utilising the exponential form of the alternating voltage and current means that, in the mathematical expressions, differentiation with respect to t can be replaced by $j\omega$.

The basic line equations have already been derived in the previous chapter as eqns. (2.2) and (2.3):

$$-\frac{\partial I}{\partial z} = GV + C\frac{\partial V}{\partial t} \quad (2.2)$$

$$-\frac{\partial V}{\partial z} = RI + L\frac{\partial I}{\partial t} \quad (2.3)$$

Making the substitution,

$$\frac{\partial}{\partial t} \equiv j\omega$$

eqns. (2.2) and (2.3) become

$$\left.\begin{array}{l}\dfrac{\partial I}{\partial z} = -(G + j\omega C)V \\[1em] \dfrac{\partial V}{\partial z} = -(R + j\omega L)I\end{array}\right\} \quad (3.1)$$

Let the total series impedance per unit length be Z; then

$$Z = R + j\omega L$$

Also let the total shunt admittance per unit length be Y; then

$$Y = G + j\omega C$$

Equation (3.1) may be written in terms of Z and Y,

$$\frac{\partial I}{\partial z} = -YV \quad (3.2)$$

$$\frac{\partial V}{\partial z} = -ZI \quad (3.3)$$

The solution to eqns. (3.2) and (3.3) is similar to that of eqns. (2.4) and (2.5); therefore

$$\frac{\partial^2 V}{\partial z^2} = YZV = \gamma^2 V \quad (3.4)$$

$$\frac{\partial^2 I}{\partial z^2} = YZI = \gamma^2 I \quad (3.5)$$

where

$$YZ = \gamma^2 \quad (3.6)$$

The solution to eqn. (3.4) is

$$V = A\exp(\gamma z) + B\exp(-\gamma z) \tag{3.7}$$

The solution to eqn. (3.5) is

$$I = C\exp(\gamma z) + D\exp(-\gamma z)$$

where A, B, C and D are arbitrary constants. But substituting from eqn. (3.7) into eqn. (3.3) gives

$$I = -\frac{\gamma}{Z}[A\exp(\gamma z) - B\exp(-\gamma z)] \tag{3.8}$$

so that C and D are not required and the arbitrary constants A and B are sufficient to supply a solution to eqns. (3.2) and (3.3). Define a new parameter taken from eqn. (3.8), and substituting for γ from eqn. (3.6) gives

$$Z_0 = \frac{Z}{\gamma} = \sqrt{\left(\frac{Z}{Y}\right)} \tag{3.9}$$

Then eqn. (3.8) becomes

$$Z_0 I = -A\exp(\gamma z) + B\exp(-\gamma z) \tag{3.10}$$

Z_0 is the same as the characteristic impedance defined earlier and in eqn. (2.13). If the conditions for no loss are applied, $R = 0$ and $G = 0$ and eqn. (3.9) reduces to eqn. (2.13). The arbitrary constants of integration are both voltage phasors and can be complex numbers. Their magnitude and phase are determined by the boundary conditions at the ends of the line. The constants Z_0 and γ are also complex numbers. Z_0 is an impedance which is mainly resistive. γ is called the *propagation constant* of the wave on the transmission line. Let the real and imaginary components of γ be α and β respectively. Then

$$\gamma = \sqrt{(YZ)} = \sqrt{[(G + j\omega C)(R + j\omega L)]} = \alpha + j\beta \tag{3.11}$$

As with eqn. (2.11), the two arbitrary constants, A and B, represent the amplitudes of a reverse wave and a forward travelling wave on the line. If eqn. (3.7) is rewritten incorporating the exponential time dependence and splitting γ into its real and imaginary parts, then

$$\begin{aligned}V &= A\exp[(\alpha + j\beta)z]\exp(j\omega t) + B\exp[-(\alpha + j\beta)z]\exp(j\omega t)\\ &= A\exp(\alpha z)\exp[j(\beta z + \omega t)] + B\exp(-\alpha z)\exp[j(-\beta z + \omega t)]\\ &= A\exp(\alpha z)\exp[j\beta(z + vt)] + B\exp(-\alpha z)\exp[-j\beta(z - vt)]\end{aligned}$$

By comparison with eqn. (2.11), it is seen that the first term represents a reverse travelling wave and the second term a forward travelling wave. v is seen to be the velocity of the wave,

$$v = \frac{\omega}{\beta} \tag{3.12}$$

Taking the real part of the complex exponentials, the voltage waveform for the forward wave is seen to be

$$V = V_0 \exp(-\alpha z) \cos[\beta(z - vt)] \tag{3.13}$$

The voltage represented by eqn. (3.13) is illustrated in Figure 3.1 for time $t = 0$ and for time $t = \pi/2\omega$. It is seen that the real part of γ occurs in the amplitude of the sine wave and the imaginary part in the period of the wave. The wavelength of the wave is the distance in which the wave shape is repeated. If λ is the wavelength, then

$$\beta\lambda = 2\pi$$

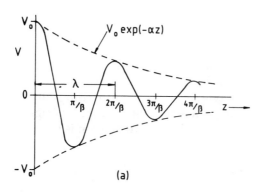

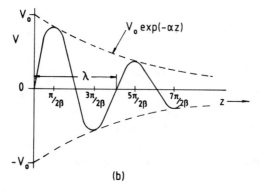

Figure 3.1 Illustration of an attenuated wave on a transmission line given by $V = V_0 \exp(-\alpha z) \cos \beta(z - vt)]$ for (a) $t = 0$, (b) $t = \pi/2\omega$.

or
$$\lambda = \frac{2\pi}{\beta} \qquad (3.14)$$

and
$$\beta = \frac{2\pi}{\lambda}$$

The voltage of the wave decreases in magnitude and is retarded in phase as one moves along the line. The reverse wave behaves in an exactly similar manner except that it is travelling in the negative z-direction.

The quantity v denotes the speed of the wave and is usually called the *phase velocity* of the wave. It is given by

$$v = \frac{\omega}{\beta} = (2\pi f)\frac{\lambda}{2\pi} = f\lambda \qquad (3.15)$$

where f is the frequency and the time period is given by

$$\tau = \frac{1}{f}$$

The rate at which the amplitude of the wave is reduced as it propagates along the transmission line is governed by the constant α. Therefore α is called the *attenuation constant*. It is measured in units of nepers/metre. If a transmission line of a given length has an input voltage amplitude V_1 and an output voltage amplitude of V_2, its attenuation is given by

$$\text{attenuation} = \ln\left|\frac{V_1}{V_2}\right| \text{ nepers}$$

In electrical circuits and systems, attenuation is more usually measured in bels or decibels. The bel is the logarithm to the base ten of the power ratio. It was adopted as a measure of attenuation of power gain because human perception of sound level appears to be logarithmic. In decibels (dB),

$$\text{attenuation} = 10\log\left|\frac{P_1}{P_2}\right| = 10\log\left|\frac{V_1^2 Z_2}{Z_1 V_2^2}\right| \text{ dB}$$

where P_1 and P_2 are the input and output power levels, respectively, and Z_1 and Z_2 are the characteristic impedances of the corresponding lines or circuits on which these powers are measured. Provided $Z_1 = Z_2 = Z_0$, which is true if the measurements are made on a uniform transmission line,

$$\text{attenuation} = 20\log\left|\frac{V_1}{V_2}\right| \text{ dB}$$

Again, provided the measurements are made on a uniform transmission line, 1 neper = 8.686 dB.

The wavelength of the wave is described by the constant β. Consequently β is called the *phase constant*. It is a measure of the relative phase of the voltage at different parts of the transmission line. Then βz is an angle measured in radians so that β is measured in radians per metre. If we write the propagation constant in terms of the properties of the transmission line as given in eqn. (3.11),

$$\begin{aligned}\gamma &= \sqrt{[(G + j\omega C)(R + j\omega L)]} \\ &= \sqrt{[-\omega^2 LC + GR + j\omega(GL + RC)]} \\ &= j\omega\sqrt{(LC)}\sqrt{\left[1 - \frac{RG}{\omega^2 LC} - j\left(\frac{G}{\omega C} + \frac{R}{\omega L}\right)\right]}\end{aligned} \quad (3.16)$$

For a low-loss line,

$$\frac{R}{\omega L} \ll 1 \quad \text{and} \quad \frac{G}{\omega C} \ll 1$$

Then eqn. (3.16) may be simplified and expanded using the first terms of a Taylor expansion of the square root,

$$\begin{aligned}\gamma &\approx j\omega\sqrt{(LC)}\sqrt{\left[1 - j\left(\frac{G}{\omega C} + \frac{R}{\omega L}\right)\right]} \\ &\approx j\omega\sqrt{(LC)}\left[1 - j\left(\frac{G}{2\omega C} + \frac{R}{2\omega L}\right)\right]\end{aligned}$$

Then by comparison with eqn. (3.11),

$$\beta = \omega\sqrt{(LC)} \quad (3.17)$$

and

$$\alpha \approx \frac{G}{2}\sqrt{\left(\frac{L}{C}\right)} + \frac{R}{2}\sqrt{\left(\frac{C}{L}\right)} = \frac{GZ_0}{2} + \frac{R}{2Z_0} \quad (3.18)$$

In eqn. (3.18), the first term is the energy loss due to the conductance between the two conductors of the transmission line, that is, the energy loss in the insulator of the line. The second term is the energy loss due to the resistance of the conductors of the line.

Equation (3.13) gives an expression for the voltage on the transmission line. The corresponding current in the line is given by

$$I = \frac{V_0}{Z_0}\exp(-\alpha z)\cos[\beta(z - vt)] \quad (3.19)$$

The phase velocity is given in eqn. (3.15),

$$v = \frac{\omega}{\beta} = \frac{1}{\sqrt{(LC)}}$$

which is the same as the wave velocity given in eqn. (2.10). Here it is seen that the propagation velocity of a voltage step on the transmission line is the same as the phase velocity of an alternating voltage on the same transmission line. If the transmission line consists of two conductors in air or vacuum as the insulation medium between them, then the phase velocity approximates to the speed of light, which is 3×10^8 m/s. The wavelength is given by eqn. (3.15). Then a voltage at 300 MHz has a wavelength of 1.0 m and a voltage at 50 Hz has a wavelength of 6×10^6 m, which is 3730 miles.

The characteristic impedance is the ratio of the amplitude of the voltage to the current on the line. It is given by eqn. (3.9), which in terms of the line constants becomes

$$Z_0 = \sqrt{\left(\frac{Z}{Y}\right)} = \sqrt{\left(\frac{R + j\omega L}{G + j\omega C}\right)} = \sqrt{\left[\frac{L}{C}\left(\frac{1 - jR/\omega L}{1 - jG/\omega C}\right)\right]} \quad (3.20)$$

Therefore for the low-loss line, we have the approximation

$$Z_0 = \sqrt{\left(\frac{L}{C}\right)}$$

which is the same as eqn. (2.13). However, for the low-frequency approximation, even for a low-loss line,

$$\frac{R}{\omega L} \gg 1 \quad \text{and} \quad \frac{G}{\omega C} \gg 1$$

and the approximate expression for Z_0 from eqn. (3.20) becomes

$$Z_0 \approx \sqrt{\left(\frac{R}{G}\right)}$$

If a line is going to transmit a wide bandwidth signal without distortion, the characteristic impedance needs to be the same at both low and high frequencies. Therefore

$$\frac{L}{C} = \frac{R}{G} \quad \text{or} \quad LG = RG \quad (3.21)$$

For most transmission lines, $LG < RC$ so that the condition of eqn. (3.21) is not satisfied. It would not be sensible to increase the loss on the line, so that G must continue to be as small as possible, but it is possible to increase L. Inductance coils can be inserted into the line at a spacing that is small compared with the wavelength on the line which provides a discrete loading of the line. Alternatively the inductance may be increased by winding a high-permeability tape round the outside of the conductor of the line, to provide a continuous loading of the line. Inductive loading not only minimises the distortion due to the line but it also minimises the signal attenuation by the line. It is left to the reader to show that the minimum of the attenuation constant as given by eqn. (3.18) occurs for the condition of eqn. (3.21). Also, substituting eqn. (3.21) into eqn. (3.18) shows that the contribution towards the attenuation due to the series resistance and that due to the shunt conductance is equal. Inductive loading is used for long-distance low-frequency transmission lines carrying communication signals.

SUMMARY

In terms of the angular frequency of an alternating voltage on a line, the line equations become

$$\frac{\partial I}{\partial z} = -(G + j\omega C)V = -YV \tag{3.2}$$

$$\frac{\partial V}{\partial z} = -(R + j\omega L)I = -ZI \tag{3.3}$$

$$\frac{\partial^2 V}{\partial z^2} = \gamma^2 V \tag{3.4}$$

$$\frac{\partial^2 I}{\partial z^2} = \gamma^2 I \tag{3.5}$$

The *propagation constant*

$$\gamma = \sqrt{(YZ)} = \sqrt{[(G + j\omega C)(R + j\omega L)]} = \alpha + j\beta \tag{3.11}$$

The solution to the wave equations is a forward and a reverse wave on the transmission line. For the forward wave only

$$V = V_0 \exp(-\alpha z) \cos[\beta(z - vt)] \tag{3.13}$$

$$Z_0 I = V_0 \exp(-\alpha z) \cos[\beta(z - vt)] \tag{3.19}$$

The *phase velocity*

$$v = \frac{\omega}{\beta} = f\lambda = \frac{1}{\sqrt{(LC)}} \tag{3.15}$$

The *phase constant*

$$\beta = \omega\sqrt{(LC)} \tag{3.17}$$

66 TRANSMISSION LINES II: A.C. EFFECTS

The *attenuation constant*

$$\alpha = \frac{GZ_0}{2} + \frac{R}{2Z_0} \qquad (3.18)$$

The *characteristic impedance*

$$Z_0 = \frac{Z}{\gamma} = \sqrt{\left(\frac{Z}{Y}\right)} = \sqrt{\left(\frac{R + j\omega L}{G + j\omega C}\right)} \approx \sqrt{\left(\frac{L}{C}\right)} \qquad (3.9), (3.20)$$

The condition for minimum signal distortion on the line is also the condition for minimum loss,

$$LG = RC \qquad (3.21)$$

The condition is effected by inductive loading of the line.

Example 3.1 Expressions for the primary line constants of a coaxial transmission line are

$$L = \frac{1}{2\pi} \mu_0 \mu_r \ln\left(\frac{b}{a}\right) \qquad \text{H/m}$$

$$C = \frac{2\pi \varepsilon_0 \varepsilon_r}{\ln(b/a)} \qquad \text{F/m}$$

where a and b are the radii of the inner and outer conductors, respectively. Find the characteristic impedance of the line and the phase velocity on the line. Show that the phase velocity is the same as the speed of light for an air-filled line where $\mu_r = 1$ and $\varepsilon_r = 1$. What is the ratio of the radii required to give a 50 Ω line for (a) air-filled insulation and (b) PTFE insulation with $\varepsilon_r = 2$, $\mu_r = 1$?

Answer The characteristic impedance is found by substituting the expressions for L and C into eqn. (3.21). Therefore

$$Z_0 = \sqrt{\left(\frac{L}{C}\right)} = \frac{1}{2\pi} \ln\left(\frac{b}{a}\right) \sqrt{\left(\frac{\mu_0 \mu_r}{\varepsilon_0 \varepsilon_r}\right)}$$

The phase velocity is given by substituting values into eqn. (3.15),

$$v = \frac{1}{\sqrt{(LC)}} = \frac{1}{\sqrt{(\mu_0 \mu_r \varepsilon_0 \varepsilon_r)}}$$

For the air-filled line,

$$v = \frac{1}{\sqrt{(\mu_0 \varepsilon_0)}} = \sqrt{\left(\frac{36\pi \times 10^9}{4\pi \times 10^{-7}}\right)} = 3 \times 10^8 \text{ m/s}$$

which is the same as the speed of light.

The expression $\sqrt{(\mu_0/\varepsilon_0)} = 120\pi$ so that the characteristic impedance of the air-filled line is given by

$$Z_0 = 60 \ln\left(\frac{b}{a}\right)$$

If the required line impedance is 50 Ω, then

$$\ln\left(\frac{b}{a}\right) = \frac{50}{60} = 0.83$$

and

$$\frac{b}{a} = 2.30$$

Therefore the required radii ratio for the air-filled coaxial line is 2.30:1. Using the PTFE insulation,

$$Z_0 = \frac{60}{\sqrt{2}} \ln\left(\frac{b}{a}\right)$$

and

$$\ln\left(\frac{b}{a}\right) = \frac{50\sqrt{2}}{60} = 1.18$$

Therefore

$$\frac{b}{a} = 3.25$$

Example 3.2 Calculate the attenuation constant for each of the lines described in Examples 2.1 and 2.6.

Answer The attenuation constant is found by substitution into eqn. (3.18). From Example 2.1 and its answer we have, $R = 6.3 \times 10^{-3}$ Ω/m, $G = 0.2 \times 10^{-9}$ S/m, $Z_0 = 685$ Ω. Therefore

$$\alpha = \frac{1}{2} \times 0.2 \times 10^{-9} \times 685 + \frac{6.3 \times 10^{-3}}{2 \times 685}$$

$$= 68.5 \times 10^{-9} + 4.60 \times 10^{-6} = 4.7 \times 10^{-6} \text{ nepers/m}$$

The contribution to the attenuation from the shunt conductance of the open-wire telephone line is about 1.5% of that due to the series resistance of the line.
From Example 2.6 and its answer we have $R = 70 \times 10^{-6}$ Ω/m, $G = 3 \times 10^{12}$ S/m, $Z_0 = 548$ Ω. Therefore

$$\alpha = \frac{1}{2} \times 3 \times 10^{-12} \times 548 + \frac{70 \times 10^{-6}}{2 \times 548}$$

$$= 822 \times 10^{-12} + 63.9 \times 10^{-9} = 65 \times 10^{-9} \text{ nepers/m}$$

68 TRANSMISSION LINES II: A.C. EFFECTS

The attenuation of the transcontinental d.c. transmission line is much less than that of the open-wire telephone line. The contribution towards the attenuation constant from the shunt conductance is still only about 1% of the total attenuation.

Example 3.3 Calculate the attenuation both in nepers and in dB and the voltage ratio between the ends of the lines for (a) 10 km of the telephone line described in Examples 2.1 and 3.2, and (b) the 180 km of the transmission line described in Examples 2.6 and 3.2.

Answer (a) The attenuation constant for the telephone line has been found in the answer to Example 3.2 to be 4.6×10^{-6} nepers/m. Therefore the attenuation due to 10 km of that line is 46×10^{-3} nepers. This is the same as 0.40 dB. The voltage ratio is given by $\exp(\alpha z)$. Therefore the required voltage ratio is $\exp(46 \times 10^{-3}) = 1.05$.

(b) The attenuation constant for the transcontinental transmission line has also been found in the answer to Example 3.2. It is 65×10^{-9} nepers/m. Therefore the attenuation due to 180 km of that line is 11.7×10^{-3} nepers. This is the same as 0.10 dB. The voltage ratio is $\exp(11.7 \times 10^{-3}) = 1.01$. This result shows that the attenuation on the transcontinential transmission line is negligible and that we were justified in the answer to Example 2.6 in neglecting any loss on the line.

PROBLEMS

3.1 For the coaxial cable described in Problem 2.1, find the attenuation constant both in nepers and decibels and the ratio of the voltages at two points 100 m apart on the line.

$[127 \times 10^{-6}$ nepers/m, 1.10×10^{-3} dB/m, 1.01$]$

3.2 For the open-wire telephone line described in Problem 2.3, find the attenuation constant and the ratio of the voltages on each end of a line 10 km long.

$[5.05 \times 10^{-3}$ nepers/km, 0.044 dB/km, 1.05$]$

3.3 The amplitude of the current in a long line terminated in its characteristic impedance, so that there is no reverse wave on the line, is measured at two points 100 m apart. The ratio of the two values is 1.1. Find the attenuation constant of the line in nepers/m. What would be the ratio of the potential differences between the two conductors of the line at two points 20 m apart?

$[9.5 \times 10^{-4}$ nepers/m, 1.02$]$

3.2 STANDING WAVES

It has already been shown in the last chapter (Section 2.1) that there can be no reflected waves from the end of an infinitely long line and that a line terminated in its characteristic impedance behaves like an infinitely long line. This is also true for an alternating voltage and current on a transmission line. Provided that the line is terminated in its characteristic impedance, there is no reverse wave on the line.

If the characteristic impedance of a line is not known, it cannot be terminated in its characteristic impedance. However, it may be determined simply by measuring its input impedance with the end of the line either open- or short-circuited. Any two-port circuit may be replaced by an equivalent-T circuit. The equivalent-T circuit of a short transmission line terminated in its characteristic impedance is shown in Figure 3.2. The input impedance of the equivalent-T circuit terminated in Z_0 is given by

$$Z_{in} = Z_1 + \frac{Z_2(Z_1 + Z_0)}{Z_1 + Z_2 + Z_0} \qquad (3.22)$$

The input impedance is also Z_0, whence

$$Z_0^2 = Z_1^2 + 2Z_1 Z_2 \qquad (3.23)$$

From the equivalent-T circuit of the line of Figure 3.2 the open-circuit input impedance is given by

$$Z_{oc} = Z_1 + Z_2 \qquad (3.24)$$

and the short-circuit input impedance by

$$Z_{sc} = Z_1 + \frac{Z_1 Z_2}{Z_1 + Z_2} = \frac{Z_1^2 + 2Z_1 Z_2}{Z_1 + Z_2} \qquad (3.25)$$

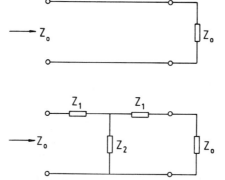

Figure 3.2 The equivalent-T circuit of a short transmission line (see p. ix).

Substitution from eqns. (3.23) and (3.24) into eqn. (3.25) gives

$$Z_0 = \sqrt{(Z_{sc} Z_{oc})} \qquad (3.26)$$

The characteristic impedance of the line is the geometric mean of the open- and short-circuit impedances.

However, if a line is not terminated in its characteristic impedance, reflected waves are set up on the line and it is necessary to consider both the forward and reverse waves in eqns. (3.7) and (3.10). Consider a line terminated in some impedance different from its characteristic impedance, as shown in Figure 3.3, where d is the distance measured from the receiving end. Then, in terms of the distance d, eqns. (3.7) and (3.10) become

$$V = A_1 \exp(-\gamma d) + B_1 \exp(\gamma d) \qquad (3.27)$$

$$Z_0 I = -A_1 \exp(-\gamma d) + B_1 \exp(\gamma d) \qquad (3.28)$$

where A_1 and B_1 are new arbitrary constants because the zero for distance has changed. When $d = 0$, $V = V_r$, $I = I_r$ and $V_r = Z_r I_r$. Therefore

$$V_r = A_1 + B_1$$

$$Z_0 I_r = -A_1 + B_1$$

Solving these simultaneous equations gives

$$B_1 = \tfrac{1}{2}(V_r + Z_0 I_r) = \tfrac{1}{2} V_r (1 + Z_0/Z_r)$$

$$A_1 = \tfrac{1}{2}(V_r - Z_0 I_r) = \tfrac{1}{2} V_r (1 - Z_0/Z_r)$$

The reflection coefficient is defined as the ratio of the voltage of the reflected wave to that of the forward wave:

$$\rho = \frac{\text{reflected voltage}}{\text{forward voltage}} = \frac{A_1}{B_1} = \frac{Z_r - Z_0}{Z_r + Z_0} \qquad (3.29)$$

which is the same result as that given in eqn. (2.23). Generally A_1 and B_1 are voltage phasors so that $\rho = |\rho| \angle \phi$ is also a complex quantity. At any

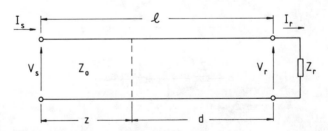

Figure 3.3 A transmission line terminated in an impedance different from its characteristic impedance.

STANDING WAVES

distance d from the receiver,

$$V = B_1[\exp(\gamma d) + \rho \exp(-\gamma d)] \tag{3.30}$$

and

$$Z_0 I = B_1[\exp(\gamma d) - \rho \exp(-\gamma d)] \tag{3.31}$$

The current reflection coefficient is given by

$$\rho_c = \frac{\text{reflected current}}{\text{forward current}} = \frac{Z_0 - Z_r}{Z_0 + Z_r} = -\rho \tag{3.32}$$

which is the same as eqn. (2.24). The current in the reflected wave is flowing in the direction of propagation of the wave which is the negative z-direction.

The sum of the reflected and forward waves on the transmission line gives rise to a standing wave on the line. Consider a low-loss line which may be considered as without loss, $\gamma = j\beta$, then eqn. (3.7) becomes

$$\begin{aligned} V &= A \exp(j\beta z) + B \exp(-j\beta z) \\ &= A(\cos \beta z + j \sin \beta z) + B(\cos \beta z - j \sin \beta z) \\ &= (A + B) \cos \beta z + j(A - B) \sin \beta z \end{aligned} \tag{3.33}$$

Equation (3.33) shows that the in-phase and antiphase components of the voltage on the line vary sinusoidally as you move down the line. Then the modulus of the voltage is given by

$$\begin{aligned} |V| &= \sqrt{[(A+B)^2 \cos^2 \beta z + (A-B)^2 \sin^2 \beta z]} \\ &= \sqrt{(A^2 + B^2 + 2AB \cos 2\beta z)} \end{aligned} \tag{3.34}$$

The amplitude of the voltage oscillates between

$$V_{max} = B + A$$

and

$$V_{min} = B - A$$

The depth of the standing wave on the line is measured as the *voltage standing wave ratio*, VSWR. It is the ratio of the voltage maximum to the voltage minimum on the line. Therefore the VSWR is given by

$$S = \frac{V_{max}}{V_{min}} = \frac{B + A}{B - A} = \frac{1 + A/B}{1 - A/B} = \frac{1 + |\rho|}{1 - |\rho|} \tag{3.35}$$

Conversely the modulus of the reflection coefficient may be obtained from a measurement of the VSWR,

$$|\rho| = \frac{S - 1}{S + 1} \tag{3.36}$$

It is seen from eqn. (3.34) that the standing wave repeats itself every half wavelength of the forward and reverse waves on the line.

There is a similar standing wave for the current on the line. For the low-loss line, eqn. (3.10) becomes

$$Z_0 I = -A \exp(j\beta z) + B \exp(-j\beta z)$$
$$= (-A + B) \cos \beta z - j(A + B) \sin \beta z \tag{3.37}$$

Then

$$Z_0 |I| = \sqrt{(A^2 + B^2 - 2AB \cos 2\beta z)} \tag{3.38}$$

The position of I_{max} coincides with the position of V_{min} and vice versa. The standing waves on lines terminated in an open circuit and a short circuit are shown in Figure 3.4.

Consider a lossless line of characteristic impedance Z_0 connected to an infinite length of another lossless line of characteristic impedance Z_r. For the

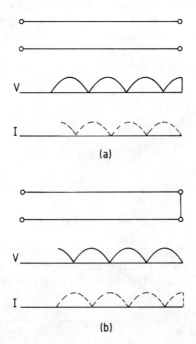

Figure 3.4 Standing waves on transmission lines: (a) open-circuit termination; (b) short-circuit termination.

lossless lines, the characteristic impedance is purely resistive, so that this is the same as a lossless line terminated in a resistance. The situation is similar to that shown in Figure 3.3, where Z_r is purely resistive. Then the modulus of ρ is given by

$$|\rho| = \frac{Z_r - Z_0}{Z_r + Z_0} \quad \text{for } Z_r > Z_0$$

and the VSWR is given by

$$S = \frac{1 + |\rho|}{1 - |\rho|} = \frac{Z_r}{Z_0} \tag{3.39}$$

However,

$$S = \frac{Z_0}{Z_r} \quad \text{for } Z_r < Z_0$$

The VSWR is given by the very simple relationship of the ratio of the two impedances.

The power carried in the transmission line is the product of the voltage and current on the line. First the voltage and current will be related to the arbitrary constants in eqn. (3.7),

$$V_{max} = A + B$$

and

$$I_{max} = \frac{A + B}{Z_0}$$

Therefore

$$V_{max} = Z_0 I_{max}$$

Similarly

$$V_{min} = Z_0 I_{min}$$

Substituting from eqn. (3.35)

$$V_{max} = S V_{min} = S Z_0 I_{min}$$

At any point on the line the power in the wave is given by

$$P = |V| \cdot |I| \cos \phi = VI^* \tag{3.40}$$

where ϕ is the power factor and the * indicates the complex conjugate. Substitution from eqns. (3.33) and (3.3) shows that

$$P = \frac{B^2}{Z_0} - \frac{A^2}{Z_0} \tag{3.41}$$

where the first term on the right-hand side of eqn. (3.41) is the power associated with the forward wave and the second term is the power associated with the reverse wave. Inspection of eqns. (3.33), (3.34) and (3.38) shows that, at the nodes and antinodes of the standing wave pattern, the voltage and current are in phase. Therefore

$$P = V_{max}I_{min} = V_{min}I_{max} = \frac{V_{max}^2}{SZ_0} = \frac{SV_{min}^2}{Z_0} = \frac{V_{max}V_{min}}{Z_0} = \frac{(A+B)(B-A)}{Z_0}$$

and the last expression reduces to eqn. (3.41).

In a lossless line the power is constant along the line. However, in a practical line, power is absorbed by the line or radiated from the line, or both, and so is lost from the line. Therefore, with reference to Figure 3.3, the efficiency of the line is defined by

$$\eta = \frac{\text{power output at } r}{\text{power input at } s} = \frac{|V_r| \cdot |I_r| \cos \phi_r}{|V_s| \cdot |I_s| \cos \phi_s} \tag{3.42}$$

where ϕ_r and ϕ_s are the power factors between the voltages and currents at the receiving and sending ends respectively.

SUMMARY

The characteristic impedance of a short length of line may be determined by measuring the input impedance of the line with the other end open- or short-circuited:

$$Z_0 = \sqrt{(Z_{sc}Z_{oc})} \tag{3.26}$$

$$\text{Reflection coefficient} = \rho = \frac{Z_r - Z_0}{Z_r + Z_0} \tag{3.29}$$

$$\text{Current reflection coefficient} = \rho_c = -\rho = \frac{Z_0 - Z_r}{Z_0 + Z_r} \tag{3.32}$$

Voltage standing wave ratio (VSWR)

$$S = \frac{1 + |\rho|}{1 - |\rho|} = \frac{V_{max}}{V_{min}} \tag{3.35}$$

$$|\rho| = \frac{S - 1}{S + 1} \tag{3.36}$$

The peak of the current standing wave coincides with the minimum of the voltage standing wave and vice versa.

For a line terminated in a resistive impedance,

$$S = \begin{cases} \dfrac{Z_r}{Z_0} & \text{for } Z_r > Z_0 \\ \dfrac{Z_0}{Z_r} & \text{for } Z_r < Z_0 \end{cases} \qquad (3.39)$$

The power on the line is the difference between the powers associated with the forward and reverse waves:

$$P = \frac{B^2}{Z_0} - \frac{A^2}{Z_0} = V_{max}I_{min} = V_{min}I_{max} = \frac{V_{max}^2}{SZ_0} = \frac{SV_{min}^2}{Z_0} \qquad (3.41)$$

The efficiency is given by

$$\eta = \frac{\text{power output at } r}{\text{power input at } s} = \frac{|V_r| \cdot |I_r| \cos \phi_r}{|V_s| \cdot |I_s| \cos \phi_s} \qquad (3.42)$$

Example 3.4 Find the reflection coefficient, VSWR and form of the voltage standing wave for a transmission line of characteristic impedance Z_0 terminated in (a) an open circuit, (b) a short circuit and (c) its matched termination Z_0.

Answer (a) For the open-circuited line, $Z_r = \infty$, $\rho = 1$, $S = \infty$ and $A = B$. Substitution into eqn. (3.33) shows that the voltage on the line is given by

$$V = 2B \cos \beta z$$

The current is given by substitution into eqn. (3.38):

$$Z_0 I = -j2B \sin \beta z$$

The form of the standing wave for both the voltage and the current is shown in Figure 3.4.

(b) For the short-circuited line, $Z_r = 0$, $\rho = -1$, $S = \infty$ and $A = -B$. The voltage and current on the line are obtained by substitution into eqns. (3.33) and (3.38):

$$V = -j2B \sin \beta z$$

$$Z_0 I = 2B \cos \beta z$$

The form of these standing waves is also shown in Figure 3.4 where it is seen that the roles of voltage and current are reversed compared with the open-circuited line.

(c) For the matched line terminated in its characteristic impedance, $Z_r = Z_0$, $\rho = 0$, $S = 1$ and $A = 0$. Substitution into eqn. (3.34) gives

$$|V| = B$$

and a plot of the voltage amplitude along the transmission line is a straight line giving a constant voltage and current amplitude.

Example 3.5 A transmission line, 100 m long, has the properties, $R = 0.26\ \Omega/\text{m}$, $L = 170\ \text{nH/m}$, $C = 66\ \text{pF/m}$, $G = 2\ \text{pS/m}$. It is connected to an oscillator with a source impedance of 50 Ω having an output of 1 V at 300 MHz. Calculate the output (voltage, current and power) delivered to a matched load at the end of the line and the efficiency of the line.

Answer The secondary constants for the line are obtained by substitution into eqns. (3.9) and (3.18):

$$Z_0 = \sqrt{\left(\frac{L}{C}\right)} = \sqrt{\left(\frac{170 \times 10^{-9}}{66 \times 10^{-12}}\right)} = 50.75$$

$$\alpha = \frac{R}{2Z_0} + \frac{1}{2}GZ_0 = \frac{0.26}{2 \times 50.75} + \frac{2 \times 10^{-12} \times 50.75}{2} = 0.0025\ \text{nepers/m}$$

If the amplitude at the source is V_0, the amplitude at the load is given by

$$V = V_0 \exp(-\alpha z)$$

If the source open circuit output is 1.0 V, the voltage connected to the line is given by dividing the available voltage between the internal impedance of the source and the characteristic impedance of the line similar to eqn. (2.25). Therefore

$$V_0 = \frac{1.0 \times 50.75}{50 + 50.75} = 0.503\ \text{V}$$

The output voltage is given by

$$V = 0.503 \exp(-0.25) = 0.392\ \text{V}$$

The output current is

$$I = \frac{V}{Z_0} = \frac{0.392}{50.75} = 7.73\ \text{mA}$$

and the output power is given by

$$P_{\text{out}} = VI = 0.392 \times 7.73 = 3.03\ \text{mW}$$

The input power is

$$P_{\text{in}} = \frac{V_0^2}{Z_0} = \frac{0.503^2}{50.75} = 5.00\ \text{mW}$$

Then the efficiency is given by

$$\eta = \frac{P_{\text{out}} \times 100}{P_{\text{in}}} = \frac{3.03 \times 100}{5.00} = 60.6\%$$

PROBLEMS

3.4 A line of characteristic impedance 50 Ω is terminated by a resistance of 100 Ω. Plot the shape of the standing wave pattern on the line.

[VSWR = 2]

3.5 A 1000 Hz generator of internal voltage 3.0 V and internal impedance of 600 Ω is connected to one end of a 240 km length of telephone line, as described in Problem 2.3. A matched termination is connected to the other end of the line. Find the voltage, current and power at the output end of the line and the efficiency of the line.

[0.466 V, 0.703 mA, 0.328 mW, 8.8%]

3.3 TERMINATING CONDITIONS

Consider a length of line terminated in an impedance Z_r as shown in Figure 3.3. It has a length l and the voltage at the source and the receiver are as shown in that diagram. The voltage and current at any point on the line are given by eqns. (3.7) and (3.10) respectively. At the source end of the line, $z = 0$, $V = V_s$ and $I = I_s$. Therefore

$$V_s = A + B$$
$$Z_0 I_s = -A + B$$

Solving for A and B gives

$$A = \tfrac{1}{2}(V_s - Z_0 I_s)$$

and

$$B = \tfrac{1}{2}(V_s + Z_0 I_s)$$

Therefore substituting into eqns. (3.7) and (3.10) gives

$$V = \tfrac{1}{2}(V_s - Z_0 I_s)\exp(\gamma z) + \tfrac{1}{2}(V_s + Z_0 I_s)\exp(-\gamma z)$$
$$Z_0 I = -\tfrac{1}{2}(V_s - Z_0 I_s)\exp(\gamma z) + \tfrac{1}{2}(V_s + Z_0 I_s)\exp(-\gamma z)$$

which may be rearranged to give

$$V = V_s \cosh \gamma z - Z_0 I_s \sinh \gamma z \qquad (3.43)$$

$$I = I_s \cosh \gamma z - \frac{V_s}{Z_0} \sinh \gamma z \qquad (3.44)$$

At the receiving end, $z = l$, $V = V_r$ and $I = I_r$. Therefore

$$V_r = V_s \cosh \gamma l - Z_0 I_s \sinh \gamma l \qquad (3.45)$$

$$I_r = I_s \cosh \gamma l - \frac{V_s}{Z_0} \sinh \gamma l \qquad (3.46)$$

If the distance is measured from the receiving end, the voltage and current on the line are given by eqns. (3.27) and (3.28) respectively. When $d = 0$, $V = V_r$ and $I = I_r$. Then eqns. (3.27) and (3.28) become

$$V = V_r \cosh \gamma d + Z_0 I_r \sinh \gamma d \qquad (3.47)$$

$$I = I_r \cosh \gamma d + \frac{V_r}{Z_0} \sinh \gamma d \qquad (3.48)$$

At the source, $d = l$, $V = V_s$ and $I = I_s$. Therefore

$$V_s = V_r \cosh \gamma l + Z_0 I_r \sinh \gamma l \qquad (3.49)$$

$$I_s = I_r \cosh \gamma l + \frac{V_r}{Z_0} \sinh \gamma l \qquad (3.50)$$

Any two-port electrical network may be represented by a black box as illustrated in Figure 3.5. The properties of the network can be expressed by a number of different network parameters. We have chosen to use the chain or ABCD network parameters which are defined by the relationships

$$\left. \begin{array}{l} V_s = AV_r + BI_r \\ I_s = CV_r + DI_r \end{array} \right\} \qquad (3.51)$$

Comparison with eqns. (3.49) and (3.50) show that the network parameters for a length of transmission line are given by

$$\left. \begin{array}{l} A = D = \cosh \gamma l \\ B = Z_0 \sinh \gamma l \\ C = \dfrac{1}{Z_0} \sinh \gamma l \end{array} \right\} \qquad (3.52)$$

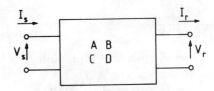

Figure 3.5 Network representation of a two-port circuit.

For a lossless line where $\gamma = j\beta$, the input impedance of a length of line l terminated in an impedance Z_r is given by the ratio of eqn. (3.49) to eqn. (3.50). Therefore

$$Z_{in} = Z_0 \frac{V_r \cos\beta l + jZ_0 I_r \sin\beta l}{Z_0 I_r \cos\beta l + jV_r \sin\beta l}$$

Dividing denominator and numerator by $I_r \cos\beta l$ gives

$$Z_{in} = Z_0 \frac{Z_r + jZ_0 \tan\beta l}{Z_0 + jZ_r \tan\beta l} \tag{3.53}$$

For measurements on a transmission line, it is often convenient if all the impedances are normalised with respect to the characteristic impedance of that line. That is, the relative or normalised value of an impedance is the absolute value of that impedance divided by the characteristic impedance of the line. Then the normalised impedance is given by

$$Z_n = \frac{Z}{Z_0}$$

However, for convenience, in the rest of this chapter all impedances will be normalised to the characteristic impedance of the transmission line in which measurements are being made and the subscript n will be omitted. Then in the normalised form eqn. (3.53) becomes

$$Z_{in} = \frac{Z_r + j\tan\beta l}{1 + jZ_r \tan\beta l} \tag{3.54}$$

which relationship enables us to find the effective impedance anywhere on a uniform transmission line due to its terminating impedance. It is seen from eqn. (3.54) that the effective impedance on the line varies between Z_r when $\tan\beta l = 0$ to $1/Z_r$ when $\tan\beta l = \infty$. If the characteristic admittance of the line is defined by $Y_0 = 1/Z_0$, all other impedances may be specified as normalised admittances. Then, in terms of admittances, it is left as an exercise for the reader to show that

$$Y_{in} = \frac{Y_r + j\tan\beta l}{1 + jY_r \tan\beta l} \tag{3.55}$$

It is also possible to obtain a relationship similar to eqn. (3.54) in terms of reflection coefficient. The input impedance of a line of length l is given by direct substitution from eqns. (3.27) and (3.28):

$$Z_{in} = \frac{B_1 \exp(\gamma l) + A_1 \exp(-\gamma l)}{B_1 \exp(\gamma l) - A_1 \exp(-\gamma l)}$$

Therefore, using eqn. (3.29) and substituting $\gamma = j\beta$ gives

$$Z_{in} = \frac{1 + \rho \exp(-j2\beta l)}{1 - \rho \exp(-j2\beta l)} \qquad (3.56)$$

If the terminating impedance is not purely resistive, any phase information may be included in the complex form of the reflection coefficient.

The performance of calculations involving eqns. (3.54), (3.55) or (3.56) is simplified by the use of graphical methods. The commonest graph is the *Smith chart* or *circle diagram* shown in Figure 3.6. Equation (3.56) may be written in the form

$$Z = \frac{1 + W}{1 - W} \qquad (3.57)$$

where W is given by

$$W = |\rho| \exp j(\phi - 2\beta l) \qquad (3.58)$$

and ϕ is the phase angle of the reflection coefficient. The transformation given in eqn. (3.57) is a transformation between the z-plane and the w-plane. It maps the rectangular coordinates in the z-plane, $Z = R + jX$, into the coordinate lines of the circle diagram shown in Figure 3.6. The resistive and reactive components of the normalised impedance are circles and segments of circles respectively. The origin of the coordinate system is at the centre of the circle diagram. A polar plot of W on the circle diagram is a plot of the reflection coefficient in magnitude and phase transferred along a length l away from the termination. The centre of the circle diagram corresponds to the condition of zero reflection coefficient. This is the perfectly matched line terminated in its characteristic impedance. For a line terminated in other than its characteristic impedance, the modulus of the reflection coefficient remains constant but the effective impedance varies with distance along the line.

A constant reflection coefficient is equivalent to a constant radius from the centre of the circle diagram. Then the locus of the effective impedance is a circle about the centre of the diagram. The input impedance of any given length of line may be read from the circle diagram provided that the relationship between the length and the angle round the diagram is known. Reference to eqn. (3.58) shows that the angle in the polar plot of reflection coefficient turns through a complete circle when $l = \frac{1}{2}\lambda$. That is, the effective impedance on the line repeats itself every half wavelength along the line. Movement along the line is equivalent to altering the phase of the reflection coefficient, which is equivalent to moving round the circle diagram at a constant radius. Therefore the outside of the circle diagram is calibrated in fractions of a wavelenth as well as in angle.

For a half-wavelength length of line, $\beta l = \pi$ and $\tan \beta l = 0$. Substituting

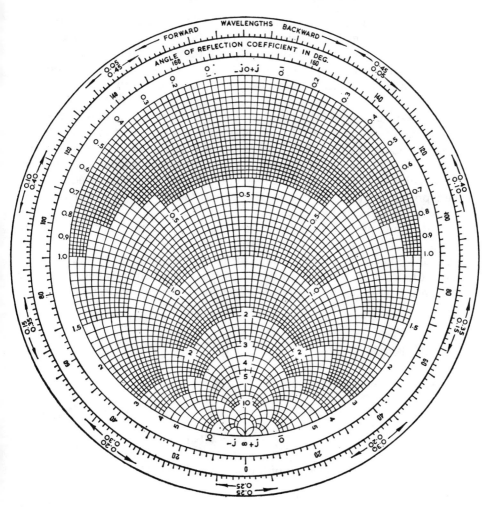

Figure 3.6 The Smith chart circle diagram (see p. ix). (Reprinted with permission from Baden Fuller, *Microwaves* (3rd end), © 1990, Pergamon Press Ltd.)

into eqn. (3.54) confirms our previous statement,

$$Z_{in} = Z_r$$

Therefore a line may be increased or decreased in length by half a wavelength or by many half wavelengths without altering its input impedance. Similarly for a quarter-wavelength of line, $\beta l = \frac{1}{2}\pi$ and $\tan \beta l = \infty$. Then

$$Z_{in} = \frac{1}{Z_r}$$

and the input impedance is the inverse of the terminating impedance. One use of this phenomenon is that an open-wire line may be supported by a pair of towers which form a transmission line a quarter wavelength high without the use of any insulators. The short circuit at the ground is transformed by the quarter-wavelength long transmission line to be an open circuit at the point of connection to the supported line. A quarter wavelength is also equivalent to moving half-way round the circle diagram, so that two points an equal distance from the centre of the circle diagram and diametrically opposite to one another denote two impedances that are the inverse of one another. If one point gives a measure of impedance, the other one gives the value of admittance of that impedance. Comparison between eqn. (3.54) and (3.55) shows that, if the circle diagram can plot the relationships of eqn. (3.54), it can also plot the relationships of eqn. (3.55). Therefore the circle diagram may also be used as an admittance diagram as well as being an impedance diagram. The impedance is $Z = R + jX$, and the circle diagram plots values for R and X. If the admittance is given by $Y = G + jB$, the circle diagram also plots values for G and B. In the impedance diagram, positive values for jX denote inductance and negative values capacitance, whereas in the admittance diagram, positive values for jB denote capacitance and negative values for jB denote inductance.

SUMMARY

The ABCD network parameters are defined by

$$\left. \begin{array}{l} V_s = AV_r + BI_r \\ I_s = CV_r + DI_r \end{array} \right\} \qquad (3.51)$$

and the parameters of a length of transmission line are

$$A = D = \cosh \gamma l, \qquad B = Z_0 \sinh \gamma l, \qquad C = \frac{1}{Z_0} \sinh \gamma l \qquad (3.52)$$

All impedances are normalised to the characteristic impedance of the transmission line, which is the same as saying that they are all measured relative to the characteristic impedance of the transmission line.

The input impedance of a length of line terminated in Z_r is

$$Z_{in} = \frac{Z_r + j \tan \beta l}{1 + j Z_r \tan \beta l} \qquad (3.54)$$

In terms of admittances,

$$Y_{in} = \frac{Y_r + j \tan \beta l}{1 + j Y_r \tan \beta l} \qquad (3.55)$$

These relationships are plotted on the *Smith chart* circle diagram.

TERMINATING CONDITIONS

The effective impedance is repeated every half wavelength along the line.
The effective impedance is inverted by a quarter wavelength of line.
The circle diagram is a polar plot of reflection coefficient.
The locus of movement along the line is a circle of constant radius about the centre of the circle diagram.

Example 3.6 For a uniform transmission line, the phase velocity is the same as the speed of light. Calculate the effective impedance at a distance of (a) 1.0 cm and (b) 2.0 cm along the line from a terminating resistance of value three times the characteristic impedance of the line. The frequency is 3.75 GHz. Confirm the results using the Smith chart.

Answer The wavelength is given by eqn. (3.15),

$$\lambda = \frac{c}{f} = \frac{3.0 \times 10^8}{3.75 \times 10^9} = 0.08 \text{ m} = 8.0 \text{ cm}$$

(a) $l = \frac{\lambda}{8}$, $\beta l = \frac{\pi}{4}$, $\tan \beta l = 1$

Substitution into eqn. (3.84) gives

$$Z_{in} = \frac{3+j}{1+3j} = 0.6 - j0.8$$

or in terms of absolute values $(0.6 - j0.8)Z_0 \, \Omega$.

(b) $l = \frac{\lambda}{4}$, $\beta l = \frac{\pi}{2}$, $\tan \beta l = \infty$

This is a half-wavelength line and the value of the impedance is inverted. Therefore $Z_{in} = \frac{1}{3}Z_0$. The graphical solutions on the Smith chart are shown in Figure 3.7.

Example 3.7 Find the ABCD network parameters of a coaxial cable at 100 MHz whose characteristic impedance is 50 Ω and phase velocity is 2.0×10^8 m/s of length (a) 0.25 m and (b) 0.50 m.

Answer The network parameters are given by eqn. (3.52). It must be assumed that the loss in the line is negligible so that $\gamma = j\beta$ and eqn. (3.52) become

$$A = D = \cos \beta l, \qquad B = jZ_0 \sin \beta l, \qquad C = \frac{j}{Z_0} \sin \beta l$$

The wavelength in the cable is given by

$$\lambda = \frac{v}{f} = \frac{2.0 \times 10^8}{1.0 \times 10^8} = 2.0 \text{ m}$$

(a) $l = \frac{\lambda}{8}$, $\beta l = \frac{\pi}{4}$, $\cos \beta l = \sin \beta l = 0.707$

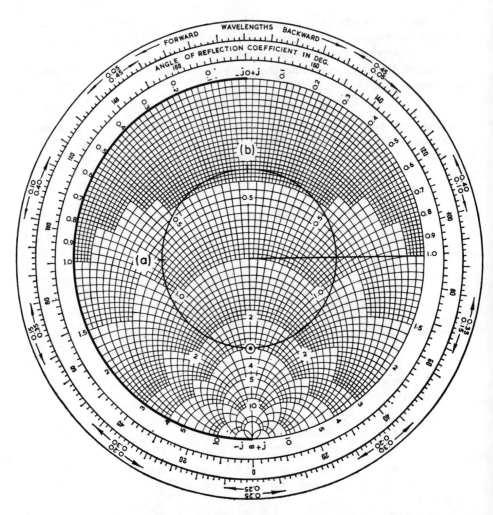

Figure 3.7 Part of the answer to Example 3.6 (a) and (b) and part of the answer to Example 3.8.

Therefore

$$A = D = 0.707, \quad B = 50 \times 0.707 = 35\,\Omega, \quad C = \frac{0.707}{50} = 0.014\,\text{S}$$

(b) $l = \dfrac{\lambda}{4}$, $\beta l = \dfrac{\pi}{2}$, $\cos \beta l = 0$, $\sin \beta l = 1$

Therefore

$$A = D = 0, \quad B = \text{j}50\,\Omega, \quad C = \text{j}0.020\,\text{S}$$

Example 3.8 What is the locus on the Smith chart circle diagram of VSWR values 1, 2, 3, 4, 5, 10 and ∞? Calculate the amplitude of the reflection coefficient appropriate to these loci.

If the termination of the transmission line is resistive and variable less than Z_0, plot on an impedance diagram the locus of the effective impedance at 0.125λ in front of the termination as its resistance value is changed.

If the termination of the transmission line is a pure reactance, inductive and variable, plot on the impedance diagram the locus of the effective impedance at 0.25λ in front of the termination as its inductance is changed.

Answer The locus on the circle diagram for any VSWR value is a circle about the centre of the diagram. The radius of the circle is determined by reference to eqn. (3.39) where it is seen that each circle passes through that point on the vertical resistive axis of the diagram where the resistance value gives the VSWR value. One VSWR circle is drawn on Figure 3.7 for a VSWR value of 3.0 and it is seen that the circle passes through the $R = 3$ point. Appropriate values for the reflection coefficient are found by substitution into eqn. (3.37). The results are given in the table below.

VSWR	1	2	3	4	5	10	∞
$|\rho|$	0	0.333	0.500	0.600	0.667	0.818	1.00

The locus of a resistive termination variable less than Z_0 is the vertical radius of the Smith chart as shown in Figure 3.6. If this is transferred along the transmission line by 0.125λ backward away from the termination, the radius is moved round 90° to the right-hand side of the diagram as shown in Figure 3.7 where it is drawn as a bold line on the diagram. The locus of a variable inductive termination is the right-hand side circumference of the circle diagram. If this is transferred along the transmission line by 0.25λ backward away from the termination, each point moves 180° around the diagram so that the locus becomes the outside circumference of the left-hand side of the circle diagram as shown by the bold line in Figure 3.7.

PROBLEMS

3.6 Find the ABCD network parameters of a transmission line which is one half-wavelength long.

[−1, 0, 0, −1]

3.7 A transmission line has a characteristic impedance of 50 Ω. Find its VSWR and effective impedance at distances of $\lambda/2$, $\lambda/4$ and $\lambda/8$ in front of a terminating impedance of (a) 50 Ω, (b) j50 Ω and (c) (75 + j25)Ω. Use the Smith chart but check three representative values analytically.

[(a) 1, 50 Ω; (b) ∞, j50, −j50, ∞ Ω; (c) 1.8, (75 + j25), (30 − j10), (60 − j30)Ω]

3.8 An air spaced transmission line is 4.0 m long and operates at 100 MHz. It has a characteristic impedance of 150 Ω. Find the input impedance for a terminating impedance of (a) (200 + j300) Ω, (b) (40 − j20) Ω and (c) (0 − j150) Ω.

[(33 + j27) Ω, (95 + j162) Ω, j555 Ω]

3.4 IMPEDANCE MEASUREMENT

The impedance connected at the end of a transmission line may be determined by measuring the amplitude and phase of the forward and reflected waves on the transmission line. Such measurement is the basis of the use of the vector voltmeter or the automatic network analyser. In a simpler and less sophisticated manner, transmission line impedances may be measured by deduction from the VSWR on the line. Reference to eqn. (3.54) and the circle diagram show that the resistive part of the effective impedance varies periodically in step with the standing-wave pattern. Therefore the minimum of the standing-wave pattern is represented by an impedance on the circle diagram located on the vertical real axis between 0 and 1. This is the position chosen for the zero of the wavelength measurement round the outside of the diagram. Reference to eqn. (3.56) shows that this is also the position for the angle π in the phase angle of the reflection coefficient. Movement round the circle diagram at a constant radius from the centre is equivalent to movement along the transmission line. The constant radius denotes a constant reflection coefficient and a constant VSWR.

The VSWR and position of the minimum enable a measurement to be made of the effective impedance at any point on the line. Consider a transmission line connected so that the only source of reflected power is from an unknown impedance at the end of the line. Then the impedance can be measured by finding the VSWR and the position of the minimum of the standing-wave pattern nearer to the generator. Let the VSWR be S and the distance between the unknown impedance and a minimum of the standing wave pattern be d. Then the effective impedance at the position of the minimum of the standing-wave pattern is known because the VSWR is at a constant radius on the diagram and the minimum is on the resistance axis between 0 and 1. On Figure 3.8, the effective impedance is shown at the point A, corresponding to $S = 3.0$. It is now necessary to move the effective impedance along the transmission line a distance $d = 0.21\lambda$ forward towards the load. The minimum of the standing-wave pattern repeats itself every half wavelength along the transmission line so that if d is greater than a half-wavelength, it is reduced by subtracting whole numbers of half wavelengths from it. The point B shows the terminating impedance to be $Z_t = 2.0 - j1.3$.

If there are a number of impedances distributed along the line, the VSWR measurement is only able to measure the total effective impedance at the point of measurement due to all the impedances contributing to the reflected wave on the line. However, if one of these impedances can be moved along the line, its contribution to the total effective impedance may easily be separated. The reflection coefficient due to each individual impedance on the line adds vectorially to the total. As the position of one impedance is moved, the phase of the reflection coefficient vector due to that impedance will change and give a circular locus on the circle diagram. The centre of the circle gives the effective impedance of all the other impedances on the line while the radius of the circle gives the modulus of the reflection coefficient

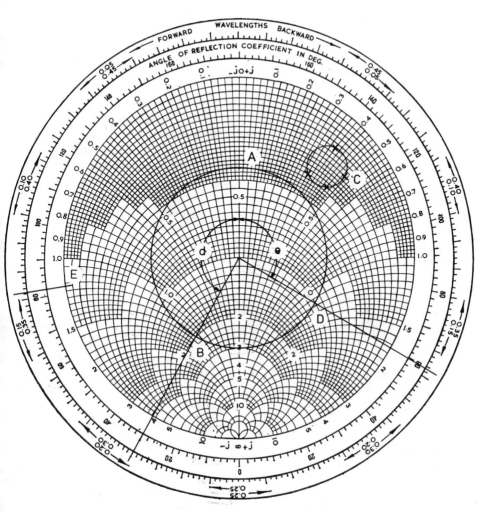

Figure 3.8 Determination of impedance from VSWR measurements. Point A is the position of the minimum. Point B is the terminating impedance. C shows the effect of using a sliding load. Point D is the position at which the termination may be matched using a series capacitance, or a shorted stub represented by the point E.

due to that one impedance. This phenomenon is most frequently used in the technique of *sliding loads*. If it is desired to measure an impedance on a transmission line which is not terminating the line, it is necessary to connect the correct characteristic impedance at the end of the line so that there is no reflected wave from the termination. If the terminating impedance is not quite the correct value, there is a small reflected wave which will cause an error in any impedance measurement. If this matched termination can move along the line, the so-called *sliding load*, its reflection coefficient may be eliminated from the measurement. The termination is moved, a number of measurements are made which define a circle on the Smith chart, and the true measurement eliminating the effect of the termination is given by the centre of the circle. A typical result is shown by the small circle drawn on Figure 3.8 labelled C. It shows the measurement of an impedance of $0.2 + j0.4$ in the presence of another impedance giving rise to a reflection coefficient of 0.115.

Another method of eliminating the effect of a number of impedances at different points on a transmission line is to use *time-domain reflectometry*. If a pulse or voltage step is impressed onto a transmission line having a number of impedances distributed on the line, a reflected pulse or voltage step is generated at each impedance. These reflected pulses return to the source at different times so that the reflection amplitude and distance along the line is known for each impedance. On a time domain reflectometer, a plot is given of relative impedance versus distance along the line.

SUMMARY

An automatic network analyser measures the forward and reverse waves on a transmission line and deduces the effective impedance on the line.

Measurement of VSWR and position of the minimum also provides the effective impedance on the line.

Use of a sliding load eliminates the effect due to the reflected wave from a poorly matched termination.

A time-domain reflectometer separates the reflection coefficients due to a number of impedances spaced on the transmission line.

Example 3.9 At a frequency of 10 GHz the terminating impedance on a transmission line was measured using a slotted line to investigate the VSWR. The results were: VSWR = 2.5, the minimum of the VSWR was 1.3 cm from the termination. The wave velocity on the line was the same as the speed of light. Find the normalised value of the terminating impedance.

Answer The wavelength is given by eqn. (3.15),

$$\lambda = \frac{c}{f} = \frac{3 \times 10^8}{10^{10}} = 0.03 \text{ m} = 3.0 \text{ cm}$$

Then the distance d from the position of the minimum to the terminating impedance is given by

$$d = \frac{1.3}{3.0} = 0.433\lambda$$

The VSWR, $S = 2.5$. These results are plotted on Figure 3.9 on the Smith chart. The position of the minimum is at A with a circle for $S = 2.5$. The terminating impedance is represented by B for a forward distance toward the load of $d = 0.433\lambda$. Then the normalised value of the terminating impedance is read off the diagram to be $Z_t = 0.47 + j0.36$.

IMPEDANCE MEASUREMENT 89

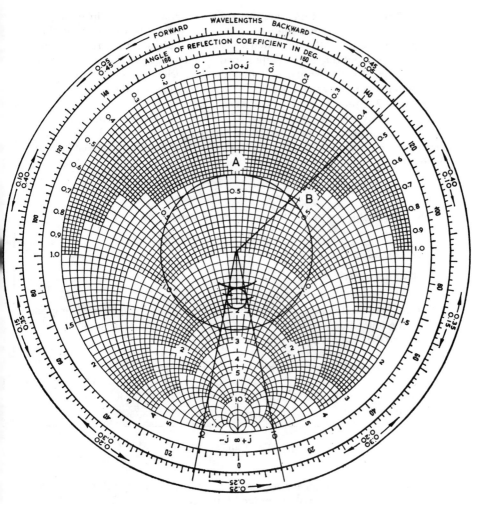

Figure 3.9 Part of the answer to Example 3.9 (points A and B) and part of the answer to Example 3.10.

Example 3.10 At a frequency of 1.5 GHz, VSWR measurements were made in an air-filled slotted line of characteristic impedance 50 Ω. The 50 Ω line was terminated in another line of unknown impedance which was then terminated in a sliding load impedance of 75 Ω. With the sliding termination in different positions, the VSWR measurements were:

VSWR	1.75	1.48	1.60
Position of minimum	0.266λ	0.250λ	0.234Ω

Find the characteristic impedance of the unknown line.

Answer The three VSWR measurements are also drawn on the circle diagram in Figure 3.9 and a circle has been drawn through them. The centre of the circle gives the impedance of the unknown line and the radius of the circle gives the reflection coefficient due to the sliding load on the unknown line. Drawing the circle is aided by the knowledge that the characteristic impedance of the unknown line is resistive and lies on the vertical real axis of the circle diagram. From Figure 3.9, the centre of the circle is at $1.65 + j0$. Therefore the impedance of the unknown line is given by $Z_1 = 1.65 \times 50 = 82.5\ \Omega$. As a cross check, the reflection coefficient is measured from the circle diagram to be 0.05. The terminating impedance is 75 Ω so that the reflection coefficient is given by eqn. (3.59):

$$\rho = \frac{Z_r - Z_0}{Z_r + Z_0} = \frac{75 - 82.5}{75 + 82.5} = -0.048$$

which agrees with the measured result within the accuracy of the measurement.

PROBLEMS

3.9 Using the Smith chart, find the impedance of the terminations giving rise to the following VSWR readings at a frequency of 3.0 GHz in an air-filled line so that $\lambda = 10$ cm.

VSWR	1.5	2.0	2.4	3.0	4.0	5.0	8.0
Position of minimum (cm)	1.25	2.50	1.88	3.68	3.08	0.63	2.13

$[(0.90 - j0.40)Z_0,\ 2.0Z_0,\ (1.0 - j1.4)Z_0,\ (0.65 + j0.85)Z_0,$
$(1.4 + j1.7)Z_0,\ (0.24 - j0.40)Z_0,\ (1.8 - j3.2)Z_0]$

3.10 A standing wave measuring line is used to measure the impedance and power in an antenna connected to its receiving end. The characteristic impedance of the line is 60 Ω. The results are VSWR = 2.2 and the first voltage minimum is located 0.37λ from the load. The value of the voltage minimum is 171 V r.m.s. Find the impedance of the antenna and the power supplied to the antenna.

$[(45 + j42)\ \Omega,\ 1.07\ \text{kW}]$

3.11 At a frequency of 300 MHz ($\lambda = 1.0$ m), slotted line measurements were made in a 50 Ω line of a junction between the 50 Ω line and another infinite line of unknown characteristic impedance. The results were: VSWR = 3, position of the minimum is 25 cm from the junction. What is the characteristic impedance of the unknown line? Given that the unknown line is now cut off at a distance of 20 cm from the junction and terminated in a pure resistance of 50 Ω, find what the new slotted line measurement will be, i.e. VSWR and position of the minimum.

$[150\ \Omega,\ 8,\ 26.6\ \text{cm}]$

3.5 IMPEDANCE MATCHING

Because the reflection coefficients due to a number of impedances distributed on a transmission line add vectorially, an impedance may be added deliberately so as to cancel any reflected waves on the line. Such a technique is called *matching*. Matching may be performed using a wide range of impedances on the line but reactive impedances are used more frequently in order to minimise any attenuation in the system. Consider the line represented by the VSWR circle on the circle diagram in Figure 3.8. Move a distance e backward from the load from the position A, the position of the minimum of the standing-wave pattern, or backward a distance $d + e$ from the the load itself to the point D on the circle diagram. This is where the constant VSWR circle crosses the $R = 1.0$ line on the graph. Then the effective impedance on the line is $(1.0 + j1.17)Z_0$. If a capacitance is now added in series with the line of value $-j1.17Z_0$, the effective impedance on the line is now Z_0 and the line is perfectly matched. The actual value of capacitance is given by $C = 1/(1.17\omega Z_0)$. From the diagram, the distance e 0.167λ and the total distance from the load is 0.377λ. D is not the only point at which matching can occur. Movement around the constant VSWR circle for another 0.166λ brings us to another point at which $R = 1.0$ and the effective impedance is now $(1.0 - j1.17)Z_0$. At this point on the line, matching is effected by adding an inductance in series with the line.

At higher frequencies, it is often difficult to realise pure inductors or capacitors and shorted lengths of line called *stubs* may be used instead of the lumped components, This gives rise to the procedure called *stub matching*. A shorted stub is a short length of transmission line with a short circuit at the end. Since the impedance of a short circuit is zero, the effective impedance of a short-circuited stub of length l is seen from eqn. (3.53) to be

$$Z = j \tan \beta l \qquad (3.59)$$

It is not necessary to perform any calculations in determining the length of shorted stubs if the rest of the design is being performed using the Smith chart because eqn. (3.59) represents the outer circumference of the Smith chart circle diagram. On this outer circumference, the resistive component of the effective impedance is zero. In the case of the shorted stub, the termination is a zero impedance represented by the point at the top outer circumference of the diagram. Then movement is backward from the load and may be read directly from the scale given round the outside of the diagram. Returning again to the transmission line represented by the VSWR circle on Figure 3.8, in order to cancel the reactive impedance at the point D, a shorted stub is required to have an effective impedance of $-j1.17Z_0$. Provided the shorted stub has the same characteristic impedance as the main transmission line, the shorted stub is represented by the point E on the circle diagram of Figure 3.8 and has a length of 0.362λ. This is not the only shorted stub that may be used to match the line. Just as an inductance may provide a match at a different point on the line, so a different length of stub can be used to provide this inductive match. It is left to the reader to show that the length of this alternative stub is 0.138λ.

In many high-frequency transmission lines, it is difficult to add stubs in series with the line so that shunt stubs are used for matching. The design procedure is similar to that used to determine the length and position of a series stub except that all the calculation is performed using the Smith chart as an *admittance diagram*. For an admittance diagram, any terminating admittance is diametrically opposite its corresponding impedance value on the diagram. Any VSWR circle is exactly the same as on the impedance diagram. The position of the minimum now corresponds to a maximum admittance so that it lies on the vertical real axis of the diagram between 1 and ∞. Distances all have to be measured from a zero at the bottom of the diagram. Similarly the length of any shorted stub is measured from a zero at the bottom of the diagram because a short circuit is an infinite admittance.

In an experimental situation as opposed to a design situation, it is often difficult to add shorted stubs exactly where they are needed, so three stubs a fixed distance apart are used. This has the advantage of allowing experimental determination of the optimum matched condition by trial and error. The theoretical design of the lengths of three matching stubs is beyond the scope of this book.

Discussion has shown how a length of transmission line may be used as an impedance-transforming device. For a purely resistive termination, a quarter wavelength of line of a particular characteristic impedance is all that is needed to match the termination. The technique is most often used to provide a match between two lines of different characteristic impedance. Consider the two lines of impedance Z_{01} and Z_{02} connected by a quarter-wavelength line of impedance Z_{03} as shown in Figure 3.10. It has already been seen from eqn. (3.54) and the Smith chart that a quarter-wavelength line transforms a terminating impedance into its inverse. Referring to Figure 3.10, the impedance at B seen relative to line 3 is

$$Z_{3B} = \frac{Z_{02}}{Z_{03}}$$

which, when transferred along line 3 to the point A, becomes Z_{03}/Z_{02}, relative to line 3. When measured from line 1 and normalised to Z_{01}, it

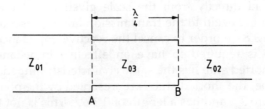

Figure 3.10 A quarter-wavelength matching section between two lines of different characteristic impedance (see p. ix).

becomes

$$Z_{1A} = \frac{Z_{03} Z_{03}}{Z_{02} Z_{01}}$$

For the matched condition, $Z_{1A} = 1$, therefore

$$Z_{03} = \sqrt{(Z_{01} Z_{02})} \qquad (3.60)$$

The junction between two lines of different characteristic impedance may be matched by inserting a quarter wavelength of line whose characteristic impedance is the geometric mean of the impedances of the lines being matched.

As will be seen from the answer to Example 3.13, matching is effective only at the one frequency at which the lengths of line are the required fractions of a wavelength. At other frequencies, they will be the wrong fractions of a wavelength. At a certain frequency, the wavelength could be such that the matching impedance increases the VSWR on the line because the mismatch due to the matching impedance adds to the mismatch already on the line rather than cancelling it out. Broadband matching requires a number of matching impedances or a number of intermediate quarter-wavelength sections of transmission line. Their design is beyond the scope of this book.

SUMMARY

Any reflected wave on a line may be cancelled by using an impedance to generate an equal and opposite reflection on the line.

At the correct position, the matching impedance is purely reactive.

A reactive matching impedance may be provided by a *shorted stub* whose impedance is given by

$$Z = j \tan \beta l \qquad (3.59)$$

Alternatively, three stubs at fixed positions may be used.

A quarter-wavelength section of line may be used to match a resistive termination or another line of different characteristic impedance. The characteristic impedance of the intermediate section must be equal to the geometric mean of the impedances of the two lines being matched.

Example 3.11 A microwave integrated circuit transistor has an input reflection coefficient relative to a 50 Ω transmission line of 0.96 ∠−40° at 4.0 GHz. What is the input impedance at 4.0 GHz? It may be matched to a 50 Ω line by connecting an inductor in series with the input and another inductor in shunt at the opposite end of the series inductor. Find the values of the two inductors.

Answer The reflection coefficient is plotted at the point A on the circle diagram of Figure 3.11. The reflection coefficient radius may be plotted directly or it may be obtained from a calculated VSWR value which can be located from the resistive scale on the diagram. For $\rho = 0.96$, $S = 1.96/0.04 = 49$. 49 is difficult to scale off the diagram but $(1/S) = 0.02$ is easy to read directly. The angle of the reflection coefficient is given round the outside of the diagram. Then from the diagram,

$$Z_{in} = (0.14 - j2.8)Z_0 = (7.0 - j140)\ \Omega$$

The matching circuit is shown in Figure 3.12. The inductance L_1 will move the effective impedance along the $R = 0.14$ line on the circle diagram to the point B. However, the location of B is determined by the calculation for L_2. Because L_2 is added in shunt, it is necessary to transfer to an admittance diagram to determine L_2. The admittance equivalent to the point B is the point C diametrically opposite to it on the circle diagram. C is located by drawing the line equivalent to the locus $R = 0.14$ on the opposite side of the diagram. The point C is located where this locus crosses the $G = 1.0$ locus on the admittance diagram.

From the point C, we have the results,

$$Y_{in} = (1.0 + j2.5)Y_0$$

L_2 is given by

$$(1/\omega L_2) = 2.5/50$$

Therefore

$$L_2 = 5.0\ \text{nH}$$

To return to the impedance diagram, the value of L_1 is given by the difference in reactance between the impedances represented by the points B and A. The point B is

$$Z = \frac{1}{Y_{in}} = (0.14 - j0.35)Z_0$$

Therefore L_1 is given by

$$\omega L_1 = (2.8 - 0.35)Z_0$$

and

$$L_1 = 30.6\ \text{nH}$$

Any real inductance of this value for use at 4 GHz would not be a pure inductance but would have a complicated equivalent circuit, so matching this transistor at this frequency is not as easy as this example indicates.

IMPEDANCE MATCHING   95

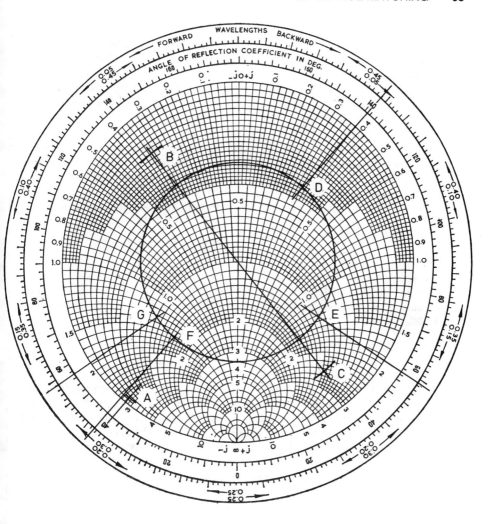

Figure 3.11 Part of the answer to Example 3.11 (points A, B and C) and part of the answer to Example 3.12 (points D, E, F and G).

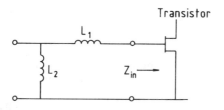

Figure 3.12 The circuit of Example 3.11.

Example 3.12 A line having a characteristic impedance of 440 Ω is terminated in an impedance of (150 + j150) Ω. It is desired to match this impedance to the line using a shorted stub. In terms of wavelength, find the length and position of the shorted stub to match the line, (a) using a series stub of characteristic impedance 330 Ω, and (b) using a shunt stub of characteristic impedance 440 Ω.

Answer (a) For a series stub an impedance diagram is used. The terminating impedance is given by $Z_r = (0.34 + j0.34)Z_0$. The solution to this problem is also drawn on the circle diagram of Figure 3.11. Z_r is represented by the point D. Then move along the constant VSWR circle backward from the load to the $R = 1.0$ circle. The distance moved is read from the outside of the diagram:

$$l_1 = (0.171 - 0.058)\lambda = 0.113\lambda$$

The effective impedance at E is $Z = (1.0 + j1.33)Z_0$. The shorted stub needs to provide $-j1.33 \times 440$ Ω, which is $-j1.77 \times 330$ Ω. The length of shorted stub to provide a normalised impedance of $-j1.77$ is read from the diagram to be 0.333λ. Therefore

$$l_2 = 0.333\lambda$$

(b) For a shunt stub, the solution requires an admittance diagram. Using the same diagram (Figure 3.11) as an admittance diagram, the admittance of the termination is diametrically opposite the impedance on the same diagram; that is, the point F on Figure 3.11. There is only a short distance to move to the $G = 1.0$ line at the point G. Then the distance l_1 is given by

$$l_1 = (0.328 - 0.307)\lambda = 0.021\lambda$$

The admittance at the point G is given by

$$Y = (1.0 - j1.34)Y_0$$

Therefore the shorted stub is required to have a normalised admittance of j1.34. The length of the shorted stub to provide this admittance is read from the diagram to be $(0.250 + 0.148)\lambda$ because the admittance of a short circuit is at the bottom of the diagram. Therefore

$$l_2 = 0.398\lambda$$

Example 3.13 Calculate the length and position of a shorted stub to be added in series with a uniform transmission line to cancel the mismatch due to a terminating impedance of $(0.5 - j0.9)Z_0$ at 300 MHz. The phase velocity is the same as the speed of light. Plot on the Smith chart the total effective impedance at the plane of the stub for a ±5% change in frequency.

IMPEDANCE MATCHING 97

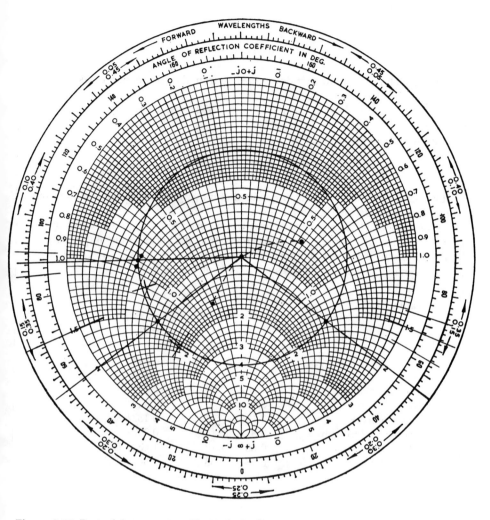

Figure 3.13 Part of the answer to Example 3.13.

Answer The construction is plotted on the circle diagram of Figure 3.13. As it is a series stub, an impedance diagram is used. The transmission line wavelength is given by

$$\lambda = \frac{c}{f} = \frac{3 \times 10^8}{3 \times 10^8} = 1.0 \text{ m}$$

The construction is drawn on the circle diagram of Figure 3.13. The results are read from the diagram to be

$$l_1 = (0.128 + 0.176)\lambda = 0.304\lambda = 30.4 \text{ cm}$$

and

$$l_2 = 0.343\lambda = 34.3 \text{ cm}$$

Alternatively, moving further down the line from the termination gives

$$l_1 = (0.128 + 0.324)\lambda = 0.452\lambda = 45.2 \text{ cm}$$

and the stub length is given by

$$l_2 = 0.157\lambda = 15.7 \text{ cm}$$

It is seen from these two sets of results that moving further down the line from the load enables a shorter stub to be used.

The second set of calculated results will be used to calculate the effect of the ±5% change in frequency. It is assumed that the terminating impedance is $R + j/\omega C$ and that there will be a similar ±5% change in ωC. The calculations are given in Table 3.1 and the results plotted on Figure 3.13.

Table 3.1

	+5%	−5%
$1/(\omega C Z_0)$	0.857	0.945
l_2/λ	0.165	0.150
X of shorted stub	$1.68 Z_0$	$1.37 Z_0$
l_1/λ	0.475	0.429
Phase of load	0.124λ	0.132λ
Transferred to stub	0.351λ	0.297λ
$(Z_t \text{ at stub})/Z_0$	$0.68 - j1.10$	$1.68 - j1.90$
$(R + jX)/Z_0$	$0.68 + j0.58$	$1.68 - j0.52$

Example 3.14 Calculate the length and impedance of the transmission lines needed to match between the 50 Ω line and the 82.5 Ω line and between the 82.5 Ω line and the 75 Ω termination of Example 3.10.

Answer Each intermediate section of line needs to be a quarter-wavelength long. The characteristic impedance is given by eqn. (3.60). For the junction between the two lines,

$$Z_{03} = \sqrt{(50 \times 82.5)} = 64.2 \text{ Ω}$$

and for the junction between the line and its termination,

$$Z_{04} = \sqrt{(82.5 \times 75)} = 78.7 \text{ Ω}$$

The completely matched system is shown in Figure 3.14.

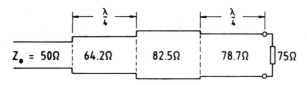

Figure 3.14 The answer to Example 3.14.

PROBLEMS

3.12 Calculate the length and position of a shorted stub to be added in series with a uniform transmission line to cancel the mismatch due to termination impedance of $(0.6 - j0.3)Z_0$ at 3.0 GHz, $\lambda = 10$ cm.

[2.15 cm, 4.07 cm; 4.15 cm 0.93 cm]

3.13 A transmission line system gives the following reading for VSWR at 3 GHz, $\lambda = 10$ cm; VSWR = 5, position of the minimum is 5.2 cm from some arbitrary datum. Find the length and position of a shorted stub to be added in shunt to cancel the mismatch of the system at 3.0 GHz.

[5.86 cm or 0.86 cm, 0.80 cm; 4.54 cm, 4.20 cm]

3.14 A 50 Ω air-spaced slotted line is used to measure an unknown terminating impedance at 1.0 GHz. The results of the measurement are VSWR = 5, the minimum is 9.0 cm from the termination. What is the absolute value of the terminating admittance? Find the length and position of a 50 Ω shunt stub to match this terminating admittance. Assume however that it is not possible to put a shunt stub in the correct design position but that a variable length shunt stub is available which can be placed 6.0 cm in front of the termination. Find the length of this fixed position shunt stub to give a minimum VSWR on the slotted line and find the slotted line measurement in this condition, i.e. VSWR and position of the minimum.

[$(4.4 - j6.4)$ mS; 2.4 cm, 7.0 cm; 3.2 cm, 1.9, 13.5 cm]

3.15 A transmission line of characteristic impedance 600 Ω is terminated in a purely resistive impedance of 1000 Ω. Find the length and characteristic impedance of an intermediate length of line to be inserted between the line and its terminating impedance so as to cancel the mismatch on the line.

[Quarter-wavelength line of impedance 775 Ω]

3.16 Design a matching section to go between two similar transmission lines having characteristic impedances of 50 Ω and 75 Ω.

[Quarter-wavelength section of 61.2 Ω]

3.17 A dipole antenna having an input impedance of 75 Ω resistive is connected to a 300 MHz source by a two-wire transmission line of characteristic impedance 200 Ω. Find the length and characteristic impedance of the two-wire air line to match the antenna to the 200 Ω line.

[25 cm, 122.5 Ω]

CHAPTER 4

Plane Waves

Aims: The aim of this chapter is to introduce the reader to electromagnetic wave propagation through media of infinite extent.

4.1 WAVE PROPAGATION

The existence of electromagnetic wave propagation is proved mathematically by the solution of Maxwell's equations in conjunction with the appropriate material properties. The particular effects of boundary conditions are considered in Chapters 5–7 in connection with guided waves. The simplest mathematical solution is obtained if boundaries are ignored and the electromagnetic fields are considered to exist in a medium of infinite extent. Maxwell's equations have already been derived as eqns. (1.26), (1.27), (1.42) and (1.52), but they are repeated here for completeness:

$$\text{div } \mathbf{D} = \nabla \cdot \mathbf{D} = \rho \tag{4.1}$$

$$\text{div } \mathbf{B} = \nabla \cdot \mathbf{B} = 0 \tag{4.2}$$

$$\text{curl } \mathbf{E} = \nabla \times \mathbf{E} = -\frac{\partial \mathbf{B}}{\partial t} \tag{4.3}$$

$$\text{curl } \mathbf{H} = \nabla \times \mathbf{H} = \mathbf{J} + \frac{\partial \mathbf{D}}{\partial t} \tag{4.4}$$

The material properties have also already been given as eqns. (1.1), (1.3) and (1.5):

$$\mathbf{D} = \varepsilon \mathbf{E} \tag{4.5}$$

$$\mathbf{B} = \mu \mathbf{H} \tag{4.6}$$

$$\mathbf{J} = \sigma \mathbf{E} \tag{4.7}$$

Initially consider propagation through a non-conducting medium. The effect of a medium of finite conductivity is considered in Section 4.6. The non-conducting medium is considered to be a perfect insulator with no stored charges. Therefore

$$\rho = 0, \quad J = 0, \quad \sigma = 0$$

in eqns. (4.1)–(4.7), and ε and μ are scalar constants. There are materials where ε and μ are dependent on field strength or vary with crystallographic direction in the material or have other additional properties, but these are beyond the scope of this book. It is assumed that all the field quantities have a sinusoidal time dependence at one frequency which can be represented by $\exp[j(\omega t)]$, so that $\partial/\partial t \equiv j\omega$, then assuming the given time dependence and substituting from eqns. (4.5) and (4.6) into eqns. (4.1)–(4.4) gives

$$\nabla \cdot \mathbf{E} = 0 \tag{4.8}$$

$$\nabla \cdot \mathbf{H} = 0 \tag{4.9}$$

$$\nabla \times \mathbf{E} = -j\omega\mu\mathbf{H} \tag{4.10}$$

$$\nabla \times \mathbf{H} = j\omega\varepsilon\mathbf{E} \tag{4.11}$$

It is now necessary to eliminate one of the two field quantities remaining in eqns. (4.8)–(4.11) to give a differential equation for the other field quantity. A similar equation is obtained whichever field quantity is eliminated. Eliminate $\mathbf{H}$ to give an equation in $\mathbf{E}$. The similar equation in $\mathbf{H}$ is quoted. It is a good exercise for the reader to perform a similar elimination to prove the correctness of eqn. (4.13). Taking the curl of both sides of eqn. (4.10), that is, operating on both sides with $\nabla\times$ and substituting from eqn. (4.11), gives

$$\nabla \times \nabla \times \mathbf{E} = -j\omega\mu\,(\nabla \times \mathbf{H}) = \omega^2\mu\varepsilon\mathbf{E}$$

The vector identity given in eqn. (1.46) is used to give an expansion for the left-hand side of this equation,

$$\nabla \times \nabla \times \mathbf{E} = \nabla(\nabla \cdot \mathbf{E}) - \nabla^2\mathbf{E}$$

and eqn. (4.8) makes the first term of this expression zero, so that

$$\nabla^2\mathbf{E} + \omega^2\mu\varepsilon\mathbf{E} = 0 \tag{4.12}$$

Eliminating $\mathbf{E}$ from eqns. (4.8)–(4.11) gives a similar expression in $\mathbf{H}$,

$$\nabla^2\mathbf{H} + \omega^2\mu\varepsilon\mathbf{H} = 0 \tag{4.13}$$

Equations (4.12) and (4.13) give a general solution to Maxwell's equations in terms of the material constants and the angular frequency of the field components. It is now necessary to use these equations to obtain solutions for the field components in terms of particular systems of space coordinates and particular physical constraints. There are two independent equations, one for **E** and one for **H**. It is possible to obtain a solution in terms of either but, when we have a solution to one, the other is related to it through Maxwell's curl equations, eqns. (4.10) and (4.11). We consider a solution for the electric field first.

In terms of the mathematics, it is found that the use of a rectangular system of coordinates is simplest. Then expanding ∇^2 in the rectangular coordinates x, y, and z, eqn. (4.12) becomes

$$\frac{\partial^2 \mathbf{E}}{\partial x^2} + \frac{\partial^2 \mathbf{E}}{\partial y^2} + \frac{\partial^2 \mathbf{E}}{\partial z^2} = -\omega^2 \mu \varepsilon \mathbf{E} \qquad (4.14)$$

In an orthogonal rectangular coordinate system, eqn. (4.14) is separable into its individual component parts, so that there is an equation similar to equation (4.14) for each of the three component parts of the vector **E**. Then for the x-directed component of the vector electric field intensity, eqn. (4.14) becomes

$$\frac{\partial^2 E_x}{\partial x^2} + \frac{\partial^2 E_x}{\partial y^2} + \frac{\partial^2 E_x}{\partial z^2} = -\omega^2 \mu \varepsilon E_x \qquad (4.15)$$

Start by finding the simplest possible solution to eqn. (4.15). As there are no boundaries to the medium in which the fields exist, the field can be uniform throughout the medium. If we decide that there is no variation of the fields in both the x- and y-directions, then

$$\frac{\partial}{\partial x} = 0 \quad \text{and} \quad \frac{\partial}{\partial y} = 0,$$

and eqn. (4.15) simplifies to

$$\frac{\partial^2 E_x}{\partial z^2} = -\omega^2 \mu \varepsilon E_x \qquad (4.16)$$

Equation (4.16) is a second-order differential equation, similar to eqn. (3.4), which has a solution (similar to eqn. (3.7))

$$E_x = A \exp(-\gamma z) + B \exp(\gamma z) \qquad (4.17)$$

where

$$\gamma^2 = -\omega^2 \mu \varepsilon \qquad (4.18)$$

It is seen that eqn. (4.17) represents a transmission-line type of wave propagating in the z-direction. E_x is one component of the field of the wave which acts in a direction perpendicular to the direction of propagation of the wave. There is no variation of the field quantities in the plane perpendicular to the direction of propagation, so the wave is called a *plane wave*.

104 PLANE WAVES

Similar to eqn. (3.6), γ is the *propagation constant* of the wave. Provided that ε and μ are both real quantities, the attenuation constant α in eqn. (3.11) is zero and the wave propagates without loss through the medium. Then

$$\gamma = j\beta$$

and the *phase constant* is given by

$$\beta = \omega\sqrt{(\mu\varepsilon)} \qquad (4.19)$$

The electromagnetic wave is travelling through the medium at a speed given by eqn. (3.12),

$$v = \frac{\omega}{\beta} = \frac{1}{\sqrt{(\mu\varepsilon)}} \qquad (4.20)$$

If the wave is propagating through a vacuum, its velocity is the speed of light,

$$v = c = \frac{1}{\sqrt{(\mu_0\,\varepsilon_0)}}$$

By definition the permeability constant is

$$\mu_0 = 4\pi \times 10^{-7} \text{ H/m}$$

The permittivity constant is related to it by the speed of light,

$$\varepsilon_0 = \frac{1}{c^2\mu_0} \text{ F/m}$$

The speed of light is

$$c = 2.998 \times 10^8 \approx 3 \times 10^8 \text{ m/s}$$

Considering only the forward wave, and incorporating the time dependence into the expression for E_x, eqn. (4.17) becomes

$$E_x = E_0 \exp[j(\omega t - \beta z)] \qquad (4.21)$$

which represents a plane wave propagating without loss through an ideal medium.

SUMMARY

In an ideal non-conducting medium, the material properties are

$$\mathbf{D} = \varepsilon \mathbf{E} \qquad (4.5)$$

$$\mathbf{B} = \mu \mathbf{H} \qquad (4.6)$$

Maxwell's equations represent the electromagnetic field components in the ideal non-conducting medium:

$$\nabla \cdot \mathbf{E} = 0 \tag{4.8}$$

$$\nabla \cdot \mathbf{B} = 0 \tag{4.9}$$

$$\nabla \times \mathbf{E} = -j\omega\mu\mathbf{H} \tag{4.10}$$

$$\nabla \times \mathbf{H} = j\omega\varepsilon\mathbf{E} \tag{4.11}$$

The solution of Maxwell's equations is the *wave equations*:

$$\nabla^2 \mathbf{E} + \omega^2 \mu\varepsilon \mathbf{E} = 0 \tag{4.12}$$

$$\nabla^2 \mathbf{H} + \omega^2 \mu\varepsilon \mathbf{H} = 0 \tag{4.13}$$

The *plane-wave* solution to eqn. (4.12) is

$$E_x = A\exp(-\gamma z) + B\exp(\gamma z) \tag{4.17}$$

The *propagation constant* is given by

$$\gamma^2 = -\omega^2 \mu\varepsilon \tag{4.18}$$

The *phase constant* is given by

$$\beta = \omega\sqrt{(\mu\varepsilon)} \tag{4.19}$$

The *velocity* is given by

$$v = \frac{1}{\sqrt{(\mu\varepsilon)}} \tag{4.20}$$

The complete expression for the forward wave incorporating the time dependence is given by

$$E_x = E_0 \exp[j(\omega t - \beta z)] \tag{4.21}$$

Example 4.1 Using Maxwell's equations in their integral form, consider the fluxes threading the faces of an elementary volume δx, δy, δz and so deduce the wave equation corresponding to the establishment of a magnetic field described by $H_y = H_z = 0$, $H_x = f(z, t)$. Express the wave equation in terms of H_x.

Answer One of the Maxwell curl equations is Faraday's law, which is given in eqn. (1.12). Applying eqn. (1.12) to one of the y–z-plane faces of the cube gives

$$\left(\frac{\partial E_z}{\partial y}\delta y\right)\delta z - \left(\frac{\partial E_y}{\partial z}\delta z\right)\delta y = -\mu_0 \frac{\partial H_x}{\partial t}\delta z \delta y$$

Therefore

$$\frac{\partial E_z}{\partial y} - \frac{\partial E_y}{\partial z} = -\mu_0 \frac{\partial H_x}{\partial t}$$

Similarly, from Ampère's law, eqn. (1.53):

$$\frac{\partial H_x}{\partial z} = \varepsilon \frac{\partial E_y}{\partial t}$$

and

$$E_z = 0$$

Therefore

$$\frac{\partial^2 H_x}{\partial z^2} = \varepsilon \frac{\partial^2 E_y}{\partial t \partial z} = \varepsilon \mu \frac{\partial^2 H_x}{\partial t^2}$$

which is the required wave equation.

PROBLEMS

4.1 Calculate the speed of an electromagnetic plane wave through the following materials each of which has relative permeability 1:

	air	polystyrene	beryllium oxide
ε_r	1	2.5	6.6

[3.0, 1.9, 1.17 × 10^8 m/s]

4.2 An electromagnetic pulse is sent from the Earth to the Moon and the reflected pulse is received 2.56 s later. How far is the Moon from the Earth?

[384 Mm]

4.2 FIELD COMPONENTS OF THE PLANE WAVE

There are two separate but similar equations which are known as the wave equation, and these are given as eqns. (4.12) and (4.13). As these are vector equations they represent three similar equations for each component of each vector field quantity. Although this represents six different wave equations, it does not represent six independent solutions to Maxwell's equations because the six different components of the two field quantities are related by the Maxwell curl equations, (4.10) and (4.11). These equations are now expanded with each vector quantity separated into three orthogonal components. Equation (4.10) becomes

$$\left. \begin{array}{l} \dfrac{\partial E_z}{\partial y} - \dfrac{\partial E_y}{\partial z} = -j\omega\mu H_x \\[6pt] \dfrac{\partial E_x}{\partial z} - \dfrac{\partial E_z}{\partial x} = -j\omega\mu H_y \\[6pt] \dfrac{\partial E_y}{\partial x} - \dfrac{\partial E_x}{\partial y} = -j\omega\mu H_z \end{array} \right\} \quad (4.22)$$

Equation (4.11) becomes

$$\left. \begin{array}{l} \dfrac{\partial H_z}{\partial y} - \dfrac{\partial H_y}{\partial z} = j\omega\varepsilon E_x \\[6pt] \dfrac{\partial H_x}{\partial z} - \dfrac{\partial H_z}{\partial x} = j\omega\varepsilon E_y \\[6pt] \dfrac{\partial H_y}{\partial x} - \dfrac{\partial H_x}{\partial y} = j\omega\varepsilon E_z \end{array} \right\} \quad (4.23)$$

The plane wave is propagating in the z-direction and there is no variation of the field quantities in the plane perpendicular to the direction of propagation. As there was the same wave equation for each component of the field quantities, they will all vary similarly to E_x as given in eqn. (4.21). Therefore the conditions for a plane wave are given by

$$\dfrac{\partial}{\partial x} = 0, \quad \dfrac{\partial}{\partial y} = 0 \quad \text{and} \quad \dfrac{\partial}{\partial z} = -j\beta$$

Substituting these conditions into eqns. (4.22) and (4.23) gives

$$\left. \begin{array}{l} j\beta E_y = -j\omega\mu H_x \\ -j\beta E_x = -j\omega\mu H_y \\ 0 = -j\omega\mu H_z \end{array} \right\} \quad (4.24)$$

$$\left. \begin{array}{l} j\beta H_y = j\omega\varepsilon E_x \\ -j\beta H_x = j\omega\varepsilon E_y \\ 0 = j\omega\varepsilon E_z \end{array} \right\} \quad (4.25)$$

Substituting for β from eqn. (4.19) and using the simplification

$$\eta = \sqrt{\left(\dfrac{\mu}{\varepsilon}\right)} \quad (4.26)$$

eqns. (4.24) and (4.25) simplify to

$$\left.\begin{array}{l} E_y = -\eta H_x \\ E_x = \eta H_y \\ E_z = 0 \\ H_z = 0 \end{array}\right\} \quad (4.27)$$

Note the simplification in eqn. (4.27). The plane wave has a very simple relationship between the different field components of the wave. If E_x exists, then only H_y needs to exist, and all the other components of the fields are zero. Alternatively, a completely independent wave may exist with the field components E_y and H_x. Because of the simplicity of the plane wave, there is no coupling between the two perpendicular components of **E** or **H**. The two perpendicularly orientated waves are completely independent and each may exist without affecting the other.

Considering the plane wave having an E_x component of its fields, the electric field component is given by eqn. (4.21) and the magnetic field component by substitution into eqn. (4.27). Then the wave is completely represented by the components

$$E_x = E_0 \exp[j(\omega t - \beta z)] \quad (4.21)$$

$$H_y = \frac{E_0}{\eta} \exp[j(\omega t - \beta z)] \quad (4.28)$$

All the other components of the field are zero. The ratio of the two field components is a constant, η, which has the dimensions of impedance and is called the *intrinsic impedance* of the medium. If the medium is a vacuum, the intrinsic impedance is given by

$$\eta_0 = \sqrt{\left(\frac{\mu_0}{\varepsilon_0}\right)} \approx 120\pi = 377 \; \Omega$$

which is sometimes called the intrinsic impedance of free space. However, there is nothing special about its particular value. It is only a property of the dimensions that are used.

If a wave is generated by a point source, the wave is uniform on a spherical wavefront. If the wave is so far from its source that the curvature of the wavefront is negligible, the wave appears to be a plane wave. A plane wave exists in space a great distance from its source and has the properties given by eqns. (4.21) and (4.28). Generally, a more complicated wave may be analysed as the sum of a number of separate plane waves all of the same frequency but having different orientations and different amplitudes and phase.

Summary of the properties of a plane wave

Our mathematical analysis has shown that a plane wave has the following characteristics:

- no fields acting in the direction of propagation;
- no variation of the field in the plane perpendicular to the direction of propagation;
- an electric field normal to the magnetic field and both act in a direction along the plane of the wave, that is in a direction perpendicular to the direction of propagation;
- the electric field is directly related to the magnetic field by the intrinsic impedance;
- the electric and magnetic fields are in time phase with one another.

A plane wave is the wave that propagates normally in unbounded free space. It starts from infinity, it goes to infinity and it extends to infinity all round. No boundaries are allowed. A plane wave cannot exist when any boundary conditions have to be considered. A plane wave exists in practice when the source is a long distance away so that any curvature of the wavefront may be neglected and when any boundaries are also a long distance away. In this context, the expressions long and large are both relative to the wavelength of the electromagnetic wave. Light may often be considered to be a plane wave in most practical situations. For radio waves, however, a true plane wave exists only for signals originating from a satellite or another planet.

The wavelength of a transmission line wave is given by eqn. (3.14); it is the distance in which the waveform repeats itself provided that time is frozen. Substituting for β from eqn. (4.19) gives the wavelength of a plane wave:

$$\lambda = \frac{2\pi}{\beta} = \frac{2\pi}{\omega\sqrt{(\mu\varepsilon)}} \qquad (4.29)$$

The wavelength of a plane wave is a function only of frequency and the material constants. If the plane wave is propagating in free space, that is, in a vacuum, then the wavelength is called the *characteristic wavelength* of the wave, and is given by

$$\lambda_0 = \frac{2\pi}{\omega\sqrt{(\mu_0\varepsilon_0)}} = \frac{2\pi c}{\omega} = \frac{c}{f} \qquad (4.30)$$

The velocity has been substituted for the material constants from eqn. (4.20) and f is the frequency. Then the relationship is

$$\lambda_0 f = c \qquad (4.31)$$

The characteristic wavelength of the wave is related to the frequency by the speed of light so it is sometimes used as a measure of the frequency of the wave. The frequencies of many broadcast stations used to be known by their characteristic wavelength. Electromagnetic waves cover the full spectrum from radio waves at low frequencies up to light and x-rays and γ-rays at high frequencies. The full electromagnetic spectrum is shown in Figure 4.1. Mathematically, the properties of all electromagnetic waves are the same whatever their frequency. However, boundary effects scale with wavelength so that the properties of electromagnetic waves of different frequencies appear to be different.

SUMMARY

For the plane wave, Maxwell's equations reduce to

$$\left.\begin{array}{l} E_y = -\eta H_x \\ E_x = \eta H_y \\ E_z = H_z = 0 \end{array}\right\} \quad (4.27)$$

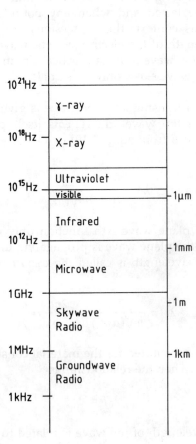

Figure 4.1 The electromagnetic spectrum.

FIELD COMPONENTS OF THE PLANE WAVE

The *intrinsic impedance*

$$\eta = \sqrt{\left(\frac{\mu}{\varepsilon}\right)} \tag{4.26}$$

A plane wave propagating in the z-direction has the fields

$$E_x = E_0 \exp[j(\omega t - \beta z)] \tag{4.21}$$

$$H_y = \frac{E_0}{\eta} \exp[j(\omega t - \beta z)] \tag{4.28}$$

The properties of a plane wave are as follows:

- no fields acting in the direction of propagation;
- no variation of field in the plane perpendicular to the direction of propagation;
- an electric field normal to the magnetic field;
- the electric field and magnetic field directly related and in time phase with one another.

The *characteristic wavelength* of the wave

$$\lambda_0 = \frac{2\pi}{\omega\sqrt{(\mu_0 \varepsilon_0)}} = \frac{c}{f} \tag{4.30}$$

$$\lambda_0 f = c \tag{4.31}$$

Example 4.2 A plane wave at 100 MHz is propagating in the z-direction. The electric field is acting in the x-direction. Its magnitude is measured to be 2.0 V/m. Find the equations for the electric and magnetic field components of the wave.

Answer As nothing is specified, it must be assumed that the wave is propagating in air or a vacuum where the velocity of the wave is the speed of light, 3.0×10^8 m/s. The wavelength of the wave is given by eqn. (4.30):

$$\lambda = \frac{3.0 \times 10^8}{1.0 \times 10^8} = 3.0 \text{ m}$$

The phase constant is given by

$$\beta = \frac{2\pi}{\lambda} = \frac{2\pi}{3.0} = 2.1 \text{ rad/m}$$

and the angular frequency is $\omega = 2\pi f = 6.28 \times 10^8$ rad/s. The peak amplitude of the electric field is $2.0 \times \sqrt{2} = 2.83$ V/m. The peak amplitude of the magnetic field is given by

$$H_y = \frac{E_0}{\eta_0} = \frac{2.38}{377} = 7.5 \times 10^{-3} \text{ A/m}$$

Then the full expressions for the field components are given by

$$E_x = 2.83 \cos (6.28 \times 10^8 t - 2.1z) \quad \text{V/m}$$
$$H_y = 7.5 \times 10^{-3} \cos (6.28 \times 10^8 t - 2.1z) \quad \text{A/m}$$

PROBLEMS

4.3 A plane wave is propagating in the z-direction. Its electric field vector makes an angle of 45° with the x-axis and has a peak amplitude of E_0. Write down expressions for the x- and y-directed components of the fields of the wave.

4.4 Calculate the characteristic wavelength compared with air of an electromagnetic plane wave in the materials given in Problem 4.1.

[0.63; 0.39]

4.3 REFLECTION AND REFRACTION FROM A PLANE BOUNDARY: (I) NORMAL INCIDENCE

We now consider what happens when a plane wave impinges on the plane boundary between two semi-infinite dielectric media, giving more evidence of the transmission line character of a plane wave. The plane boundary exists between two non-conducting dielectric media. The mathematical theory considers the ideal situation where half of space consists of medium 1 having the properties μ_1, ε_1 and η_1, and the other half of space consists of medium 2 with the properties μ_2, ε_2 and η_2. The direction of propagation of the incident wave is perpendicular to the plane of the boundary. For a plane incident wave, the fields of the various waves generated by the boundary are shown in Figure 4.2. The electric fields of all the waves are parallel, as are also the magnetic fields of all the waves. The incident wave has the fields

$$E_1 \quad \text{and} \quad H_1, \quad \text{where} \quad E_1 = \eta_1 H_1$$

which gives rise to a transmitted wave having the fields

$$E_2 \quad \text{and} \quad H_2, \quad \text{where} \quad E_2 = \eta_2 H_2$$

and a reflected wave

$$E_3 \quad \text{and} \quad H_3, \quad \text{where} \quad E_3 = -\eta_1 H_3$$

The negative sign occurs in this relationship because positive H always acts in one direction and the reflected wave is propagating in the reverse direction.

REFLECTION AND REFRACTION FROM A PLANE BOUNDARY: (I)

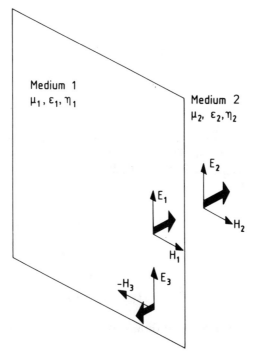

Figure 4.2 A plane boundary between two semi-infinite media showing the material properties of the two media and the field components of the forward, transmitted and reflected plane waves.

The tangential components of the fields are the same on each side of the boundary. Therefore the total field on one side of the boundary is the same as that on the other side of the boundary. Equating the fields gives

$$E_1 + E_3 = E_2 \tag{4.32}$$

and

$$H_1 + H_3 = H_2 \tag{4.33}$$

Substituting the relationships between E and H into eqn. (4.32) gives

$$\eta_1 H_1 - \eta_1 H_3 = \eta_2 H_2 \tag{4.34}$$

Solving eqns. (4.33) and (4.34) gives

$$2\eta_1 H_1 = (\eta_1 + \eta_2) H_2 \tag{4.35}$$
$$2\eta_1 H_3 = (\eta_1 - \eta_2) H_2$$

Therefore

$$\frac{H_3}{H_1} = \frac{\eta_1 - \eta_2}{\eta_1 + \eta_2} \tag{4.36}$$

and

$$\frac{E_3}{E_1} = \frac{\eta_2 - \eta_1}{\eta_2 + \eta_1} \tag{4.37}$$

Equation (4.37) gives an expression for the reflection coefficient due to the boundary. Reference to eqn. (3.29) shows that there is a similarity between the intrinsic impedance of the different media and the characteristic impedance of a transmission line. From eqn. (4.35), and similarly to eqn. (2.24), a transmission coefficient may be specified:

$$\mathcal{T} = \frac{E_2}{E_1} = \frac{2\eta_2}{\eta_1 + \eta_2} \tag{4.38}$$

and

$$\mathcal{T}_h = \frac{H_2}{H_1} = \frac{2\eta_1}{\eta_1 + \eta_2}$$

The reflected wave causes a standing wave in the medium 1 where

$$E_{max} = E_1 + E_3$$
$$E_{min} = E_1 - E_3$$

and the *voltage standing wave ratio* (VSWR) is given by

$$S = \frac{E_{max}}{E_{min}} \tag{4.39}$$

Substitution of suitable values from eqn. (4.37) gives a result similar to eqn. (3.39):

$$S = \frac{\eta_2}{\eta_1} \tag{4.40}$$

The VSWR is equal to the ratio of the intrinsic impedances in the two media in exactly the same way as the VSWR due to a junction between two transmission lines is equal to the ratio of the characteristic impedances of the two lines. Here we see further evidence that a plane wave propagating through free space behaves exactly like a transmission line wave.

In Section 3.5 it was shown from transmission line theory that a quarter wavelength of line of intermediate impedance may be inserted between two lines of different characteristic impedance in order to match the junction. Similarly, a quarter wavelength of material of intermediate intrinsic impedance may be inserted at the plane boundary between two dielectric materials in order to match the junction. The reflection coefficient is zero. The intrinsic impedance of the intermediate material is given by a relationship similar to eqn. (3.60),

$$\eta_3 = \sqrt{(\eta_1 \eta_2)} \qquad (4.41)$$

This is the theory of the blooming of lenses in optical systems where the lens is coated with a thin film of dielectric material in order to reduce reflections from the surface of the lens.

Since it is very unlikely that any of the dielectric materials involved are magnetic, it is safe to assume that

$$\mu_1 = \mu_2 = \mu_3 = \mu_0$$

and eqn. (4.41) gives an expression for the permittivity of the intermediate material:

$$\varepsilon_3 = \sqrt{(\varepsilon_1 \varepsilon_2)}$$

A quarter wavelength of absorbing material operates similarly to provide a matched termination for the plane wave, when it is coated onto a perfectly reflecting surface. The wave is neither reflected nor transmitted. It is useful in radar installations to mask nearby fixed installations, whose echo would swamp useful signals from the same direction.

SUMMARY

A plane wave normally incident on a plane boundary experiences a *reflection coefficient*,

$$\rho = \frac{\eta_2 - \eta_1}{\eta_2 + \eta_1} \qquad (4.37)$$

and a *transmission coefficient*,

$$\mathcal{T} = \frac{2\eta_1}{\eta_1 + \eta_2} \qquad (4.38)$$

The *VSWR* is

$$S = \frac{\eta_2}{\eta_1} \qquad (4.40)$$

The boundary may be matched with a quarter wavelength of material whose intrinsic impedance is

$$\eta_3 = \sqrt{(\eta_1 \eta_2)} \qquad (4.41)$$

Example 4.3 Calculate the reflection coefficient and VSWR in the air space for a plane wave in air normally incident onto the plane surface of a dielectric medium whose relative permittivity is 2.5.

Answer The reflection coefficient and VSWR are given by eqns. (4.37) and (4.40) respectively as the ratio of intrinsic impedances. The intrinsic impedance in air is the same as that in free space; therefore

$$\eta_1 = \eta_0 = 377 \ \Omega$$

The intrinsic impedance in the dielectric medium is given by

$$\eta_2 = \frac{\eta_0}{\sqrt{\varepsilon_r}} = \frac{377}{1.58} \ \Omega$$

Then the reflection coefficient is

$$\rho = \frac{1 - 1.58}{1 + 1.58} = -0.225$$

The VSWR is always greater than one; therefore in this case, the VSWR in the air space is

$$S = 1.58$$

PROBLEMS

4.5 A uniform plane wave in air impinges normally onto the plane surface of a dielectric material. The E field in the air is measured to have an SWR of 5.3 and a distance between adjacent minima of 2.0 m. What is the relative permittivity of the dielectric material and the frequency of the plane wave?

[5.3; 75 MHz]

4.6 A lens is made of glass having a permittivity of 2.1. Calculate the permittivity of the film material that would be used to bloom the lens.

[1.45]

4.4 REFLECTION AND REFRACTION FROM A PLANE BOUNDARY: (II) OBLIQUE INCIDENCE

The plane boundary between two semi-infinite dielectric media is the same as that considered in the last section. The plane wave impinges onto the surface at an angle ψ_1. A section perpendicular to the surface and including the incident ray for the plane wave is shown in Figure 4.3. The condition is

REFLECTION AND REFRACTION FROM A PLANE BOUNDARY: (II)

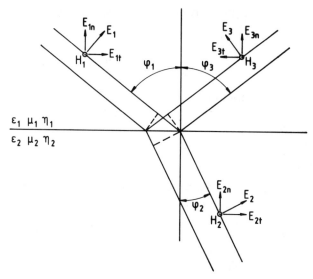

Figure 4.3 Incident, reflected and transmitted waves at the plane boundary between two semi-infinite dielectric media.

shown on the diagram with the incident plane wave orientated with the magnetic field vector parallel to the surface. Other orientations are equally possible and are considered later. The diagram shows an incident wave E_1 and H_1, a transmitted wave E_2 and H_2 and a reflected wave E_3 and H_3. The electric field vector is further shown resolved into its components parallel to the surface, with the subscript t, and perpendicular to the surface with the subscript n.

Because a plane wave is being considered which has the same properties everywhere inside each medium, there is a phase front travelling parallel to the boundary which is the same on each side of the boundary. Therefore the component of velocity parallel to the boundary for each of the waves is the same, giving

$$\frac{v_1}{\sin \psi_1} = \frac{v_1}{\sin \psi_3} = \frac{v_2}{\sin \psi_2} \tag{4.42}$$

The velocities of the incident and reflected waves are the same because they are both propagating through the same medium. Therefore

$$\psi_1 = \psi_3 \tag{4.43}$$

giving the usual law of light that the *angle of reflection equals the angle of incidence*. Further deduction from eqn. (4.42) gives Snell's law:

$$\frac{v_1}{v_2} = \frac{\sin \psi_1}{\sin \psi_2} \tag{4.44}$$

118 PLANE WAVES

but the ratio of the velocities gives the *index of refraction* between the two media:

$$n_{12} = \frac{v_1}{v_2} = \sqrt{\left(\frac{\mu_2 \varepsilon_2}{\mu_1 \varepsilon_1}\right)} \tag{4.45}$$

where the velocity is obtained from eqn. (4.20). For two non-magnetic dielectric media, the index of refraction is given by

$$n_{12} = \sqrt{\left(\frac{\varepsilon_2}{\varepsilon_1}\right)}$$

The relationships given in eqns. (4.43) and (4.44) are independent of the orientation of the plane wave incident on the surface and they are true for any plane wave impinging at oblique incidence onto a plane surface. However, the relative amplitudes of the transmitted and reflected waves are a function of the orientation of the incident plane wave. First a wave with the magnetic field parallel to the surface, as shown in Figure 4.3, is considered. In order to preserve continuity, the magnetic fields of the transmitted and reflected waves also have to be parallel to the surface. At the boundary, the tangential components of the field intensities are equal on each side of the boundary surface. It is then necessary to resolve each electric field vector into its components parallel and perpendicular to the surface. The field vectors of each plane wave are related by the material properties of the media:

$$E_1 = \eta_1 H_1, \qquad E_3 = -\eta_1 H_3, \qquad E_2 = \eta_2 H_2$$

The boundary equations are similar to eqns. (4.32) and (4.33):

$$E_{1t} + E_{3t} = E_{2t} \tag{4.46}$$

$$H_1 + H_3 = H_2 \tag{4.33}$$

By resolving and by substituting the material properties, the tangential components of the electric field are given by

$$E_{1t} = E_1 \cos \psi_1 = \eta_1 H_1 \cos \psi_1$$
$$E_{3t} = E_3 \cos \psi_1 = -\eta_1 H_3 \cos \psi_1 \tag{4.47}$$
$$E_{2t} = E_2 \cos \psi_2 = \eta_2 H_2 \cos \psi_2$$

Substituting from eqn. (4.47) into eqn. (4.46) gives

$$\eta_1 H_1 \cos \psi_1 - \eta_1 H_3 \cos \psi_1 = \eta_2 H_2 \cos \psi_2 \tag{4.48}$$

Solving the eqns. (4.33) and (4.48) to eliminate H_2 gives

$$\eta_1 \cos \psi_1 (H_1 - H_3) = \eta_2 \cos \psi_2 (H_1 + H_3)$$

from which the *reflection coefficient* is obtained:

$$\rho = \frac{H_3}{H_1} = \frac{\eta_1 \cos \psi_1 - \eta_2 \cos \psi_2}{\eta_1 \cos \psi_1 + \eta_2 \cos \psi_2} \qquad (4.49)$$

At one particular angle of incidence, the numerator of eqn. (4.49) is zero, H_3 is zero and there will be no reflected wave. The condition for no reflection is given by putting the numerator of eqn. (4.49) equal to zero; therefore

$$\frac{\cos \psi_1}{\cos \psi_2} = \frac{\eta_2}{\eta_1} = \sqrt{\left(\frac{\mu_2 \varepsilon_1}{\varepsilon_2 \mu_1}\right)} \qquad (4.50)$$

ψ_2 may be eliminated from eqns. (4.44) and (4.50) to give the value of ψ_1 for no reflection. However, for non-magnetic dielectric materials, the value for this angle is very simple. When $\mu_1 = \mu_2 = \mu_0$, eqn. (4.50) becomes

$$\cos \psi_2 = n_{12} \cos \psi_1 \qquad (4.51)$$

Taking eqns. (4.44) and (4.45) and squaring gives

$$\sin^2 \psi_1 = n_{12}^2 \sin^2 \psi_2 = n_{12}^2 (1 - \cos^2 \psi_2) = n_{12}^2 (1 - n_{12}^2 \cos^2 \psi_1)$$

Dividing through by $\cos^2 \psi_1$ gives

$$\tan^2 \psi_1 = n_{12}^2 (\sec^2 \psi_1 - n_{12}^2) = n_{12}^2 (1 + \tan^2 \psi_1 - n_{12}^2)$$

Therefore

$$\tan \psi_1 = n_{12} \qquad (4.52)$$

The angle of incidence given by eqn. (4.52) is called the *Brewster angle* and often denoted by the subscript B, ψ_{1B}.

We obtain the transmission coefficient if H_3 is eliminated between eqns. (4.33) and (4.48), giving

$$\eta_1 H_1 \cos \psi_1 - \eta_1 (H_2 - H_1) \cos \psi_1 = \eta_2 H_2 \cos \psi_2$$

whence

$$\mathcal{T}_h = \frac{H_2}{H_1} = \frac{2\eta_1 \cos \psi_1}{\eta_1 \cos \psi_1 + \eta_2 \cos \psi_2} \qquad (4.53)$$

but it is normal to define the transmission coefficient as the ratio of the electric field quantities. Therefore

$$\mathcal{T} = \frac{E_2}{E_1} = \frac{\eta_2 H_2}{\eta_1 H_1} = \frac{\eta_2}{\eta_1} \mathcal{T}_h = \frac{2\eta_2 \cos \psi_1}{\eta_1 \cos \psi_1 + \eta_2 \cos \psi_2} \qquad (4.54)$$

The mathematical deviation of expressions for the reflection coefficient and transmission coefficient when the electric field vector of the incident wave is orientated parallel to the boundary surface is very similar to eqns. (4.46)–(4.54). A summary is given here. The boundary equations are similar to eqns. (4.32) and (4.33):

$$E_1 + E_3 = E_2 \tag{4.32}$$

$$H_{1t} + H_{3t} = H_{2t} \tag{4.55}$$

The expressions for the tangential components of the magnetic field vectors are similar to eqn. (4.47); therefore eqn. (4.55) becomes

$$\frac{E_1}{\eta_1}\cos\psi_1 - \frac{E_3}{\eta_1}\cos\psi_1 = \frac{E_2}{\eta_2}\cos\psi_2 \tag{4.56}$$

E_2 is eliminated between eqns. (4.32) and (4.56), giving the reflection coefficient

$$\rho = \frac{E_3}{E_1} = \frac{\eta_2\cos\psi_1 - \eta_1\cos\psi_2}{\eta_2\cos\psi_1 + \eta_1\cos\psi_2} \tag{4.57}$$

The numerator of eqn. (4.57) cannot be zero because such a condition is incompatible with Snell's law, eqn. (4.44). Therefore there is no angle of incidence for which there is no reflected wave. The Brewster effect occurs only for the magnetic field orientated parallel to the boundary surface.

If E_3 is eliminated between eqns. (4.32) and (4.56), the transmission coefficient is obtained:

$$\mathcal{T} = \frac{E_2}{E_1} = \frac{2\eta_2\cos\psi_1}{\eta_2\cos\psi_1 + \eta_1\cos\psi_2} \tag{4.58}$$

If the wave is passing across the boundary from a material of larger permittivity to one of smaller permittivity, there is a range of ψ_1 such that according to Snell's law, eqn. (4.44), $\sin\psi_2$ ought to be greater than one and there can be no transmitted wave. Under those conditions, the reflection coefficient is unity, giving the condition of *total internal reflection*. The limiting condition is given by

$$\sin\psi_{1C} = n_{12} \tag{4.59}$$

However, in order to satisfy the boundary conditions, it is not sufficient just to let $E_2 = 0$. There must be some electromagnetic fields in the medium 2. Snell's law is satisfied by having complex values for ψ_2. Let

$$\psi_2 = \theta + j\phi$$

and use the identities

$$\sin\psi_2 = \sin(\theta + j\phi) = \sin\theta\cosh\phi + j\cos\theta\sinh\phi$$
$$\cos\psi_2 = \cos(\theta + j\phi) = \cos\theta\cosh\phi - j\sin\theta\sinh\phi$$

For $\sin \psi_2$ real and greater than unity, $\cos \theta = 0$ and ψ_2 is given by

$$\psi_2 = \frac{\pi}{2} + j\phi$$

Then

$$\sin \psi_2 = \cosh \phi \quad \text{and} \quad \cos \psi_2 = -j \sinh \phi$$

Substituting from Snell's law, eqn. (4.44) and (4.45) and remembering that $\varepsilon_1 > \varepsilon_2$ gives

$$\sin \psi_2 = \cosh \phi = n_{21} \sin \psi_1 \qquad (4.60)$$
$$\cos \psi_2 = -j\sqrt{(\cosh^2 \phi - 1)} = -j\sqrt{(n_{21}^2 \sin^2 \psi_1 - 1)} \qquad (4.61)$$

The theory in this section has shown that the orientation of the field vectors in a plane wave is significant. The orientation is defined by the *polarisation* of the wave. The polarisation denotes the direction of the electric field vector. The *plane of polarisation* is defined as that plane containing the electric field vector and the direction of propagation. The plane of polarisation of the plane wave given in eqns. (4.21) and (4.28) is the x–z plane. Most terrestrial radio waves are polarised in relation to the surface of the Earth. They are *horizontally* or *vertically* polarised. For lower frequency radio receivers using a ferrite rod aerial; the ferrite rod is orientated parallel to the magnetic field, that is perpendicular to the plane of polarisation of the wave. For higher-frequency receivers using a metal dipole aerial, the aerial rods are orientated parallel to electric field, that is, they are orientated into the plane of polarisation.

In the case of radio waves, the direction of polarisation of the wave is determined by the orientation of the transmitting aerial, but for light, most of the generators produce random polarisation. The Brewster effect may be used to eliminate those components of the wave that are polarised with the magnetic field parallel to the surface. There are also some dielectric materials which selectively absorb those components of the wave polarised in a particular direction. Because most surfaces are horizontal, the Brewster angle effect causes most reflected sunlight to be polarised in one direction so that the use of polarised sunglasses reduces glare. However, internal stresses in toughened glass also tend to polarise light so that polarised sunglasses cannot be worn in some vehicles with toughened glass windscreens.

SUMMARY

For a plane wave impinging onto a plane boundary at an oblique angle of incidence, *the angle of reflection equals the angle of incidence*.
Snell's law gives the angle of the transmitted wave,

$$\frac{\sin \psi_1}{\sin \psi_2} = \sqrt{\left(\frac{\mu_2 \varepsilon_2}{\mu_1 \varepsilon_1}\right)} = n_{12} \qquad (4.44), (4.45)$$

122 PLANE WAVES

When the magnetic field component of the incident wave is parallel to the boundary surface,

$$\rho = \frac{\eta_1 \cos \psi_1 - \eta_2 \cos \psi_2}{\eta_1 \cos \psi_1 + \eta_2 \cos \psi_2} \tag{4.49}$$

$$\mathcal{T} = \frac{2\eta_2 \cos \psi_1}{\eta_1 \cos \psi_1 + \eta_2 \cos \psi_2} \tag{4.54}$$

At the *Brewster angle* there is no reflected wave,

$$\tan \psi_{1B} = n_{12} \tag{4.52}$$

When the electric field component for the incident wave is parallel to the boundary surface,

$$\rho = \frac{\eta_2 \cos \psi_1 - \eta_1 \cos \psi_2}{\eta_2 \cos \psi_1 + \eta_1 \cos \psi_2} \tag{4.57}$$

$$\mathcal{T} = \frac{2\eta_2 \cos \psi_1}{\eta_2 \cos \psi_1 + \eta_1 \cos \psi_2} \tag{4.58}$$

For this condition, ρ cannot be zero.

If the wave is passing across the boundary from a material of larger permittivity to one of smaller permittivity there is a range of incident angles for which no wave is transmitted and there is *total internal reflection*.

$$\sin \psi_1 > \sin \psi_{1C} = n_{12} \tag{4.59}$$

Polarisation defines the orientation of a wave.

Plane of polarisation is that plane containing the electric field vector and the direction of propagation.

Example 4.4 A plane wave in air is incident onto a plane surface of glass having a relative permittivity of 2.1 at the Brewster angle. The plane of polarisation of the incident wave is at an angle of 45° to the plane perpendicular to the surface containing the direction of propagation of the wave. Find the polarisations and relative strengths of the reflected and transmitted waves.

Answer The direction of propagation is given by Snell's law and the Brewster angle. The refractive index of glass is given by

$$n_{12} = \sqrt{\varepsilon_r} = \sqrt{2.1} = 1.45$$

As the plane of polarisation is at an angle of 45° to a plane perpendicular to the surface, it is necessary to resolve the incident wave into two equal components, one having its electric field vector parallel to the boundary surface and one having its magnetic field vector parallel to the surface. If E_0 is the peak amplitude of the incident wave, each component of the wave has an amplitude $E_0/\sqrt{2}$. From the Brewster angle, eqn. (4.52), we obtain

$$\tan \psi_1 = n_{12} = 1.45, \quad \text{therefore} \quad \psi_1 = 55.4°$$

and

$$\sin \psi_1 = 0.82, \quad \cos \psi_1 = 0.57$$

From Snell's law, eqns. (4.44) and (4.45),

$$\sin \psi_2 = \frac{1}{n_{12}} \sin \psi_1 = 0.57 \quad \text{and} \quad \cos \psi_2 = 0.82$$

The intrinsic impedances of the two media are given by

$$\eta_1 = \eta_0 = 377 \,\Omega, \quad \eta_2 = \frac{\eta_0}{n_{12}} = \frac{\eta_0}{1.45} = \eta_0 \times 0.69$$

For that component of the wave having its magnetic field parallel to the surface, the reflection coefficient is given by eqn. (4.49):

$$\rho_H = 0$$

because the wave is incident at the Brewster angle. The transmission coefficient is given by eqn. (4.54)

$$\mathcal{T}_H = \frac{2 \times 0.57 \times 0.69}{0.57 + 0.82 \times 0.69} = 0.69$$

For that component of the wave having its electric field parallel to the surface, the reflection coefficient is given by eqn. (4.57)

$$\rho_E = \frac{0.57 \times 0.69 - 0.82}{0.57 \times 0.69 + 0.82} = -0.35$$

and the transmission coefficient by eqn. (4.58)

$$\mathcal{T}_E = \frac{2 \times 0.57 \times 0.69}{0.57 \times 0.69 + 0.82} = 0.65$$

Therefore the reflected wave is polarised in a plane perpendicular to the reference plane specified in the question. Its relative amplitude is $0.35/\sqrt{2} = 0.25$ of the incident wave. The transmitted wave is polarised at an angle given by

$$\tan^{-1} \frac{0.65}{0.69} = 43.3°$$

to the reference plane. Its relative amplitude is given by

$$\sqrt{\left(\frac{0.69^2 + 0.65^2}{2} \right)} = \frac{0.95}{\sqrt{2}} = 0.67$$

so that the transmitted wave has almost the same polarization as the incident wave.

PROBLEMS

4.7 The electric field vector of a circularly polarised plane wave rotates about an axis along the direction of propagation and traces out a helix in space if time is frozen. It may be described by the equation $E_x = jE_y$. Write expressions for all components of the fields of this wave and describe it as the sum of a number of normal plane waves.

4.8 A plane wave in air is incident onto the plane surface of a large block of dielectric having a relative permittivity of 3.25. The direction of propagation of the incident wave makes an angle of 30° with the plane of the surface, and the wave is polarised with its electric field vector parallel to the surface. Calculate the relative magnitudes and directions of the electric field intensities of the reflected and transmitted waves.

[0.52, 30°; 0.48, 61.3°]

4.5 ATTENUATION

So far in this chapter it has been assumed that the plane wave is propagating through an infinite medium without loss. The attenuation constant has been assumed to be zero. In practice this is not true except for propagation through a perfect vacuum. However, for propagation through air and through many low-loss dielectric materials it is a very good approximation. In this section, we discuss how the small loss may be accounted for. Equation (3.11) shows how, in a transmission line wave, the propagation constant may be divided into its real and imaginary parts, giving the phase constant and the attenuation constant,

$$\gamma = \alpha + j\beta \tag{4.62}$$

and eqn. (4.21) becomes

$$E_x = E_0 \exp(-\alpha z) \exp[j(\omega t - \beta z)] \tag{4.63}$$

Losses occur because the medium absorbs power from the wave as it propagates and the power heats the medium. If the losses are large, both the phase constant and the attenuation constant are affected by the loss, but if the losses are small it is satisfactory to calculate the phase constant ignoring the loss and calculate the attenuation constant separately as with eqn. (3.18). The attenuating properties of the medium may be described mathematically by specifying complex values for the permeability and permittivity;

$$\mu = \mu' - j\mu''$$
$$\varepsilon = \varepsilon' - j\varepsilon''$$

If the concept of complex values for the material properties appears to be novel, the reader is recommended to calculate the impedance of a parallel plate capacitor whose dielectric is represented by a complex permittivity. It will be found that the imaginary part of the complex permittivity gives rise to a resistive component in the impedance of the capacitor. Because the losses are usually small, they are often specified by *loss tangents*

$$\tan \delta_e = \frac{\varepsilon''}{\varepsilon'}$$

$$\tan \delta_m = \frac{\mu''}{\mu'}$$

The propagation constant is given by eqn. (4.18); therefore

$$\gamma^2 = -\omega^2(\mu' - j\mu'')(\varepsilon' - j\varepsilon'') \tag{4.64}$$

It is possible to find expressions for the real and imaginary parts of γ but they are long and complicated and will not be given here. Some results for the much simpler condition of zero magnetic loss, $\mu'' = 0$, are given in the next section as eqns. (4.72) and (4.73). For the low-loss condition,

$$\varepsilon'' \ll \varepsilon' \quad \text{and} \quad \mu'' \ll \mu'$$

and eqn. (4.64) simplifies to

$$\gamma^2 = -\omega^2 \mu' \varepsilon' [1 - j(\tan \delta_m + \tan \delta_e)] \tag{4.65}$$

When the losses are small, it is assumed that the phase constant is given by

$$\beta^2 = \omega^2 \mu' \varepsilon'$$

and substituting into eqn. (4.65) gives

$$\gamma^2 = -\beta^2[1 - j(\tan \delta_m + \tan \delta_e)]$$

Then γ can be expanded by the binomial theorem to give

$$\gamma \approx j\beta + \tfrac{1}{2}\beta(\tan \delta_m + \tan \delta_e)$$

If the intrinsic impedance of the medium is defined from the real parts of the permeability and permittivity,

$$\eta = \sqrt{\left(\frac{\mu'}{\varepsilon'}\right)}$$

the phase constant and the attenuation constant can be specified in terms of the material properties:

$$\beta = \omega\sqrt{(\mu'\varepsilon')} \qquad (4.66)$$

$$\alpha \approx \tfrac{1}{2}\beta(\tan\delta_m + \tan\delta_e) = \frac{\omega}{2}\left(\frac{\mu''}{\eta} + \varepsilon''\eta\right) \qquad (4.67)$$

The two terms in eqn. (4.67) represent the contributions from the magnetic and the dielectric loss mechanisms, respectively, and the equation is similar to eqn. (3.18). Values of the loss tangents are given in tables of material data.

Throughout the electromagnetic spectrum, different materials are transparent or are absorbing or opaque at different frequencies. Glass is a common household object which is transparent at optical frequencies but opaque at the lower infrared frequencies. Therefore a glass screen may be used in front of an open fire to shield occupants from direct heat while still allowing the fire to remain visible as a focus of attention. A greenhouse uses the same property of glass. Energy is received into the greenhouse in the form of visible and ultra-violet radiation where it is absorbed. However, the contents of the greenhouse only give out radiation at infrared frequencies which are absorbed by the glass, and consequently retain the energy inside the greenhouse.

SUMMARY

The attenuating properties of a material are specified by complex values for the permeability and permittivity

$$\mu = \mu' - j\mu'' \qquad \text{and} \qquad \varepsilon = \varepsilon' - j\varepsilon''$$

and loss tangents

$$\tan\delta_m = \frac{\mu''}{\mu'} \qquad \text{and} \qquad \tan\delta_e = \frac{\varepsilon''}{\varepsilon'}$$

For low-loss material,

$$\beta = \omega\sqrt{(\mu'\varepsilon')} \qquad (4.66)$$

$$\alpha = \tfrac{1}{2}\beta(\tan\delta_m + \tan\delta_e) = \frac{\omega}{2}\left(\frac{\mu''}{\eta} + \varepsilon''\eta\right) \qquad (4.67)$$

Example 4.5 A 10 GHz plane wave is propagating through a block of polystyrene having the dielectric properties

$$\varepsilon_r = 2.5, \qquad \tan\delta_e = 0.001, \qquad \mu_r = 1$$

Calculate the phase constant, the wavelength and the attenuation constant in decibels for the plane wave.

The same plane wave then propagates through a block of ferrite material having the properties

$$\mu_r = 10 - j0.1, \qquad \varepsilon_r = 13$$

Calculate the same quantities.

Answer Substituting from eqn. (4.20), eqn. (4.66) becomes

$$\beta = \frac{\omega}{c}\sqrt{\varepsilon_r} = \frac{2\pi 10^{10}}{3 \times 10^8}\sqrt{2.5} = 331 \text{ rad/m}$$

The wavelength is given by eqn. (4.29):

$$\lambda = \frac{2\pi}{\beta} = \frac{c}{f\sqrt{\varepsilon_r}} = 19 \text{ mm}$$

The attenuation constant is obtained from eqn. (4.67):

$$\alpha = \tfrac{1}{2}\beta \tan \delta_e = \tfrac{1}{2} \times 331 \times 0.001 = 0.166 \text{ nepers/m} = 1.44 \text{ dB/m}$$

Similar substitutions for the material properties of the ferrite material give

$$\beta = \frac{2\pi 10^{10}}{3 \times 10^8}\sqrt{130} = 2388 \text{ rad/m}$$

$$\lambda = 2.6 \text{ mm}$$

$$\alpha = \tfrac{1}{2} \times 2388 \times 0.01 = 11.9 \text{ nepers/m} = 104 \text{ dB/m}$$

PROBLEMS

4.9 A dielectric material has a relative permeability of unity and a complex relative permittivity of $2.2 - j0.25$ measured at visible frequencies. A light wave with a characteristic wavelength of 1.3 μm is propagating through the material. Calculate the phase constant and the attenuation constant of the wave.

[7.17×10^6 rad/m, 3.53×10^6 dB/m]

4.10 A non-magnetic material has a relative permittivity of 4 and a loss tangent of 0.1 at low frequencies. Calculate the wavelength and attenuation constant for a plane wave in the material at 2.5 MHz.

[60 m, 0.0455 dB/m]

4.6 CONDUCTING MEDIUM

So far in this chapter, it has been assumed that the medium through which the plane wave is propagating is a perfect insulator with no stored charges. A perfect conductor acts as a perfect reflector for electromagnetic waves. At the surface of a perfect conductor, the electric field intensity is zero and no electric field can exist inside a perfect conductor. A perfect conductor has infinite conductivity. However, real conductors have a finite conductivity and an electromagnetic field can exist inside them. In many situations it is sufficient to assume that a good conductor behaves as if it were a perfect conductor but there are situations where it is necessary to consider the real conductivity of the material.

In a medium of finite conductivity, eqn. (4.7) applies. Substituting into eqn. (4.4) gives

$$\nabla \times \mathbf{H} = \mathbf{J} + \frac{\partial \mathbf{D}}{\partial t} = \sigma \mathbf{E} + j\omega\varepsilon \mathbf{E} = (\sigma + j\omega\varepsilon)\mathbf{E} \qquad (4.68)$$

The conductivity may be combined into part of an effective complex permittivity

$$\varepsilon_{\text{eff}} = \varepsilon - j\frac{\sigma}{\omega} \qquad (4.69)$$

The effective permittivity may then be substituted for the permittivity in eqn. (4.11) and the wave equation governing the fields in the conducting medium is the same as eqns. (4.12) and (4.13). For the conducting medium, it is more convenient to work in terms of the magnetic field. Then, similar to eqn. (4.21) a solution to eqn. (4.13) is

$$H_y = H_0 \exp(j\omega t - \gamma z) \qquad (4.70)$$

where

$$\gamma^2 = -\omega^2 \mu \left(\varepsilon - j\frac{\sigma}{\omega}\right) \qquad (4.71)$$

As in eqn. (3.11), the propagation constant may be divided into its real and imaginary parts

$$\gamma = \alpha + j\beta$$

Therefore

$$\gamma^2 = \alpha^2 - \beta^2 + j2\alpha\beta = -\omega^2\mu\varepsilon + j\omega\mu\sigma$$

Equating the real and imaginary parts of this equation gives

$$\alpha^2 - \beta^2 = -\omega^2\mu\varepsilon$$

and
$$\alpha\beta = \tfrac{1}{2}\omega\mu\sigma$$

Therefore

$$\alpha^2 = \tfrac{1}{2}\omega^2\mu\varepsilon\left(-1 + \sqrt{\left[1 + \left(\frac{\sigma}{\omega\varepsilon}\right)^2\right]}\right) \quad (4.72)$$

$$\beta^2 = \tfrac{1}{2}\omega^2\mu\varepsilon\left(1 + \sqrt{\left[1 + \left(\frac{\sigma}{\omega\varepsilon}\right)^2\right]}\right) \quad (4.73)$$

As with other lossy material considered in the last section, the propagation constant has both real and imaginary parts and as the wave propagates its amplitude decays exponentially. The power lost to the electromagnetic wave goes into conduction current and heats up the material. In order to simplify the mathematics, a limitation to a material which is a good conductor is accepted. Therefore, for a good conductor,

$$\sigma \gg \omega\varepsilon$$

and eqn. (4.71) becomes

$$\gamma^2 = j\omega\mu\sigma$$

Therefore

$$\gamma = (1 + j)\sqrt{\left(\frac{\omega\mu\sigma}{2}\right)}$$

and

$$\alpha = \beta = \sqrt{(\tfrac{1}{2}\omega\mu\sigma)} \quad (4.74)$$

It has been assumed that there is a plane wave propagating through a conducting medium and the propagation properties are given by eqns. (4.72) and (4.73) or (4.74). Because it is a plane wave, a relationship similar to eqn. (4.24) also exists:

$$\gamma E_x = j\omega\mu H_y \quad (4.75)$$

H_y is given by eqn. (4.70) which substituting for γ gives

$$H_y = H_0 \exp(-\alpha z)\exp[j(\omega t - \alpha z)] \quad (4.76)$$

and therefore E_x is given by

$$E_x = \frac{j\omega\mu}{(1 + j)}\sqrt{\left(\frac{2}{\omega\mu\sigma}\right)} H_0 \exp(-\alpha z)\exp[j(\omega t - \alpha z)]$$

or

$$E_x = (1+j)\sqrt{\left(\frac{\omega\mu}{2\sigma}\right)} H_0 \exp(-\alpha z) \exp[j(\omega t - \alpha z)] \qquad (4.77)$$

The plane wave described by eqns. (4.76) and (4.77) has its field components in the plane of the wave and perpendicular to one another but the two field components are 45° *out of phase* with each other. Their relative amplitude remains constant. This plane wave is propagating through the conducting medium with a phase constant that is a function of the conductivity and whose amplitude is decaying rapidly with distance. The amplitude drops to $1/e$ in a distance $\lambda/2\pi$. In one wavelength, the amplitude drops by more than 1/500, so that it is impossible to consider a practical situation which approximates to an infinite conducting medium. However, it is possible to consider the plane surface of an infinite conducting medium.

Assume that a normally incident plane wave creates a field given by

$$H_y = H_0 \exp(j\omega t)$$

at the surface of a semi-infinite block of conducting material. The reflected wave in the air space adjacent to the conducting surface is very large but it is only the magnetic field parallel to the surface at the surface which is needed to determine the fields inside the surface. Considering the boundary conditions at the surface, the magnetic field in the conducting medium at the surface is also

$$H_y = H_0 \exp(j\omega t)$$

assuming that the surface is in the x–y plane and that the wave is propagating in the z-direction. A section through the surface is shown in Figure 4.4.

To find the total electric current flowing in the conducting medium due to the electromagnetic fields, it is necessary to integrate the current in the direction of propagation of the plane wave. Consider a unit square of the surface of the conducting medium as shown in Figure 4.4. The current is represented by the current density which is a vector parallel to the surface,

$$J_x = \sigma E_x$$

The current flowing in a small element of unit width is

$$\delta I = \sigma E_x \delta z$$

which is integrated to give the total current flowing through a unit length of the surface,

$$I_0 = \int_0^\infty \sigma E_x \, dz$$

$$= (1+j)\sqrt{\left(\frac{\omega\mu\sigma}{2}\right)} H_0 \int_0^\infty \exp[-(1+j)\alpha z] \, dz = H_0 \qquad (4.78)$$

CONDUCTING MEDIUM

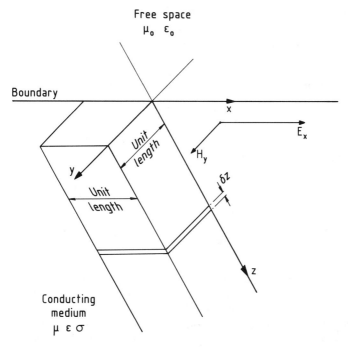

Figure 4.4 The plane boundary between a semi-infinite block of conducting material and free space.

I_0 is the total current flowing across each unit length of the surface and it is equal to the magnetic field intensity which has been applied to the surface. As the field intensity decays so rapidly with penetration into the surface, most of this current is flowing in a thin skin adjacent to the surface.

As the electromagnetic fields penetrate into the conducting medium, the amplitude of the fields decrease and power is lost in heating the material. Because an electric current is generated in the material, the power lost to the material may be calculated from the resistive heating of the medium due to the current. The power loss density is given by

$$p = \tfrac{1}{2} E_x J_x^* = \tfrac{1}{2} \sigma E_x E_x^*$$

$$= \tfrac{1}{2} \sigma (1 + j) \sqrt{\left(\frac{\omega \mu}{2\sigma}\right)} H_0 \exp[-\alpha(1 + j)z]$$

$$\times (1 - j) \sqrt{\left(\frac{\omega \mu}{2\sigma}\right)} H_0 \exp[-\alpha(1 - j)z]$$

$$= \tfrac{1}{2} \omega \mu H_0^2 \exp(-2\alpha z)$$

The power loss in a unit area thin slice as shown in Figure 4.4 is given by

$$\delta P = \tfrac{1}{2} \omega \mu H_0^2 \exp(-2\alpha z) \delta z$$

Therefore the total power loss in the unit area of infinite depth is

$$P = \int_0^\infty \tfrac{1}{2}\omega\mu H_0^2 \exp(-2\alpha z)\,dz = \frac{1}{2}\sqrt{\left(\frac{\omega\mu}{2\sigma}\right)} H_0^2 \qquad (4.79)$$

Substituting from eqn. (4.78) and rearranging gives

$$P = \frac{1}{2}\sqrt{\left(\frac{\omega\mu\sigma}{2}\right)}\frac{1}{\sigma}I_0^2 = \frac{1}{2}\frac{1}{z_0\sigma}I_0^2 \qquad (4.80)$$

If the power loss is put into the form

$$P = \tfrac{1}{2}RI_0^2$$

then

$$R = \frac{1}{z_0\sigma}$$

R is the equivalent surface resistivity of the block of conducting material and is measured in the units of $\Omega/\square$. It has the same numerical value for a square of any size of the surface. z_0 is called the *skin depth*. It is seen from eqn. (4.80) that exactly the same power would be dissipated in the medium if the same total current were equally distributed in the thickness z_0. The skin depth is also that distance in which the current and the amplitude of the electromagnetic fields drop to $1/e$ of its value at the surface. By $4z_0$ the current density is 2% of its value at the surface and it may be considered negligible. The skin depth is

$$z_0 = \frac{1}{\alpha} = \sqrt{\left(\frac{2}{\omega\mu\sigma}\right)} \qquad (4.81)$$

At any frequency, if the cross-section dimension of a conductor is greater than the skin depth, the a.c. resistance of the conductor is greater than the d.c. resistance of the same conductor. Although the theory of this section has been developed in terms of an infinite plane surface, current concentration in a thin skin at the surface of a conductor occurs whatever the shape of the conductor and the concept of skin depth still applies. For non-planar surfaces, the skin depth is not the same as that of the plane surface but the value given in eqn. (4.81) is always a good approximation. The wavelength of propagation into the surface is given by

$$\lambda = 2\pi z_0$$

Provided the radius of curvature of the surface is much greater than the wavelength and therefore much greater than the skin depth, the plane surface approximation for the calculation of skin depth is valid. The concept of current and field concentration at the surface is useful at many different frequencies. At low frequencies it is used to calculate the thickness required for the laminations of transformer cores so that they are thin compared with the skin depth to give a good penetration of the magnetic field. At higher frequencies the skin depth gives an idea of the a.c. resistance of conductor wires. At very high frequencies, the skin depth may be used to calculate the thickness of metallisation required on a half-silvered mirror.

The boundary conditions at the surface of a conductor have been given by eqn. (1.18) to be

$$H_{t1} = H_{t2} + I_s$$

where I_s is the surface current, which is the same as I_0 calculated in eqn. (4.78). For the good conductor, the surface current is dominant and it is assumed that H_{t2} is zero. We have calculated expressions for the fields in the conductor and the skin depth. In effect, the wave is reflected from the surface of the conductor and the penetration of the fields is small. The conductivity can be a function of frequency in the same way that the absorption of lossy materials can be a function of frequency. As an example, semi-conducting glass, which is perfectly transparent to light, is used as a screen against r.f. radiation.

SUMMARY

The *effective permittivity* of a conducting medium is given by

$$\varepsilon_{eff} = \varepsilon - j\frac{\sigma}{\omega} \tag{4.69}$$

The propagation constant of a plane wave in the medium

$$\gamma^2 = -\omega^2\mu\left(\varepsilon - j\frac{\sigma}{\omega}\right) \tag{4.71}$$

For a good conductor, $\sigma \gg \omega\varepsilon$ and

$$\alpha = \beta = \sqrt{(\tfrac{1}{2}\omega\mu\sigma)} \tag{4.74}$$

The components of a plane wave are

$$H_y = H_0 \exp(-\alpha z) \exp[j(\omega t - \alpha z)] \tag{4.76}$$

$$E_x = (1+j)\sqrt{\left(\frac{\omega\mu}{2\sigma}\right)} H_0 \exp(-\alpha x) \exp[j(\omega t - \alpha z)] \tag{4.77}$$

The total current density flowing parallel to a plane surface is

$$I_0 = H_0 \tag{4.78}$$

where H_0 is the amplitude of the magnetic field intensity parallel to the surface.

$$\text{Power loss density} = \frac{1}{2}\frac{1}{z_0\sigma}I_0^2 \quad (4.80)$$

$$\text{Skin depth} = z_0 = \frac{1}{\alpha} = \sqrt{\left(\frac{2}{\omega\mu\sigma}\right)} \quad (4.81)$$

Example 4.6 Copper has the following properties: $\mu = \mu_0$, $\varepsilon = \varepsilon_0$ and $\sigma = 5 \times 10^7$ S/m. Calculate the characteristic wavelength, the skin depth and the plane wave wavelength in copper at 1 MHz and 10 GHz.

Answer The characteristic wavelength is given by eqn. (4.30), the skin depth by eqn. (4.81) and the wavelength in copper by $2\pi z_0$. Inserting numbers into these equations gives

$$\lambda_0 = \frac{c}{f} = \frac{3 \times 10^8}{1 \times 10^6} = 300 \text{ m} \quad \text{for } f = 1 \text{ MHz}$$

$$= \frac{3 \times 10^8}{1 \times 10^{10}} = 3 \text{ cm} \quad \text{for } f = 10 \text{ GHz}$$

$$z_0 = \sqrt{\left(\frac{2}{\omega\mu\sigma}\right)} = \sqrt{\left(\frac{2}{2\pi \times f \times 4\pi \times 10^{-7} \times 5 \times 10^7}\right)}$$

$$= \frac{1}{2\pi}\sqrt{\left(\frac{1}{5f}\right)} = 0.071\,18\,\frac{1}{\sqrt{f}}$$

$$= \frac{0.072}{10^3} = 72\ \mu\text{m} \quad \text{for } f = 1 \text{ MHz}$$

$$= \frac{0.072}{10^5} = 0.72\ \mu\text{m} \quad \text{for } f = 10 \text{ GHz}$$

$$\lambda = 2\pi z_0 = 0.447\,\frac{1}{\sqrt{f}} = \frac{0.447}{10^3} = 447\ \mu\text{m} \quad \text{for } f = 1 \text{ MHz}$$

$$= \frac{0.447}{10^5} = 4.47\ \mu\text{m} \quad \text{for } f = 10 \text{ GHz}.$$

PROBLEMS

4.11 Using Gauss's law, Ohm's law and the equation of continuity, show that if a charge density ρ_0 could exist uniformly within a conductor at time $t = 0$, then this charge would decrease exponentially with a time constant ε/σ s. Calculate the time constant for a copper conductor. What happens to the charge?

[1.5×10^{-19} s]

4.12 Calculate the lowest frequency at which a plane surface approximation would be valid for a copper wire of 5 mm diameter.

[100 kHz]

CHAPTER 5

Waveguide Effects

Aims: This chapter gives an introductory description of the properties of waveguide propagation and a description of a number of high-frequency transmission lines and waveguides.

5.1 PARALLEL PLATE WAVEGUIDE

In the last chapter, the mathematics and the propagation characteristics of a plane wave were developed. However, there are very many situations where boundaries cannot be ignored and a plane-wave approximation is not a valid description of the wave propagation. It is all a question of scale. If the space is large compared with the characteristic wavelength of the wave, a plane-wave approximation is a valid description of the wave. If the space is of the same order as the characteristic wavelength, waveguide effects occur and these are described in this chapter. If the space is small compared to the characteristic wavelength, a.c. circuit theory and transmission line theory as described in Chapters 2 and 3 are applicable. At low radio frequencies, radio receivers are designed on the basis of a.c. circuit theory. At higher radio frequencies, such as those used for v.h.f. f.m. radio and for television broadcasting, transmission-line phenomena affect the performance of the radio receiver. At microwave frequencies with a characteristic wavelength of human dimensions, a plane-wave description is usually not valid and waveguide phenomena predominate. At light frequencies, the plane-wave description of the wave is valid. In order to use waveguide effects for light, it is necessary to use hair-thin fibres of glass as the transmission medium.

In this chapter, the general properties of waveguide as a transmission medium are developed. At microwave frequencies, hollow metal pipe is used as waveguide and is described in Section 5.4 and Chapter 6. Glass fibre is used as waveguide for light and is described in Chapter 7. In order to provide a communication link, propagation through space is one of the major uses of electromagnetic radiation. However, it is also necessary to control the electromagnetic radiation at the transmitter and the receiver and to channel it from one point to another without allowing it to escape as radiation into the surrounding space. Any system of multiple conductor transmission line controls and guides the radiation with varying amounts of containment. For complete containment, however, the transmission line needs to be enclosed inside a solid conducting screen. At microwave frequencies, hollow metal pipe is one of the most convenient totally enclosed systems and provides the lowest attenuation. Chapter 6 is devoted to the solution of Maxwell's equations inside a hollow metal pipe and the determination of the necessary conditions for propagation to occur. First, in this chapter, propagation in parallel-plate waveguide is described by means of a pictorial description of the superposition of plane waves that gives a good visualisation of waveguide propagation.

Consider the effect of putting two parallel perfectly conducting plates perpendicular to the electric field of a plane wave as shown in Figure 5.1. Because the conducting plates are perpendicular to the electric field, they do not disturb the field pattern of the wave. Therefore we can make a transmission line of the two parallel plates having a portion of the field of a plane wave between them. The wave has all the transmission line properties of the plane wave including the same wavelength. The electric field acts perpendicular to the plates and the magnetic field parallel to them. Provided the curvature is large, the plates may be bent around corners, keeping their separation constant, so guiding the plane wave. To a first approximation, the wave between the plates remains a plan wave. Such a device may be used to guide the electromagnetic wave and it is called a *waveguide*.

However, there are other possible configurations for the fields inside the waveguide which give rise to different modes of propagation. These other modes propagate with a different wavelength and are called waveguide modes. The plane wave in the parallel-plate waveguide propagates with a wavelength which is the same as the plane-wave wavelength or the characteristic wavelength of the wave and this is called the transmission-line mode.

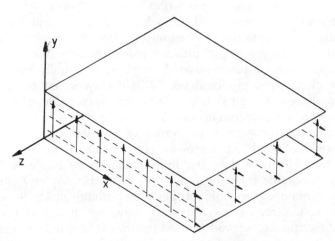

Figure 5.1 A portion of an infinite parallel-plate waveguide. ———→ electric field; -----→ magnetic field.

PARALLEL PLATE WAVEGUIDE

Now to develop the properties of the waveguide modes. Consider a plane wave reflected at an oblique angle from a plane perfectly conducting surface. The diagram in Figure 5.2 shows a ray in the direction of propagation and lines of equal phase in the wave which are perpendicular to the ray. If a second plane conducting surface is placed at AA, or at any position parallel to the other surface through points where the two sets of equal phase lines cross, the field pattern of the fields between the two surfaces is unchanged. The field outside the conducting surfaces may be ignored. The field is confined to the space between the surfaces and the ray is reflected without loss at an angle θ from the surfaces as shown in Figure 5.3.

The plane wave is propagating along the line of the ray so that the spacing of the lines of equal phase is the characteristic wavelength, λ_0, of the wave. The construction shown in Figure 5.3 is dependent on there being a formal relationship between the spacing between the conductors and the angle and wavelength of the plane wave. For this condition, it is seen that the spacing

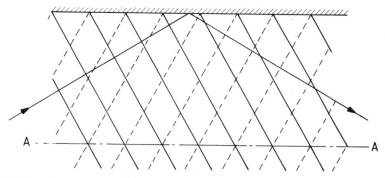

Figure 5.2 A plane wave reflected from a plane conducting sheet, showing a ray in the direction of propagation and lines of equal phase for both the incident and reflected wave (see p. ix).

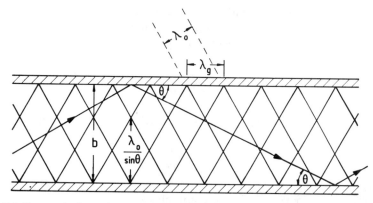

Figure 5.3 Ray and phase front representation of an electromagnetic wave propagating between two parallel conducting surfaces. One ray is shown and it is perpendicular to the lines of equal phase. The separation between the plates is b. The waveguide wavelength and the plane-wave wavelength are also shown (see p. ix).

between the conductors is given by

$$b = \frac{n\lambda_0}{2\sin\theta} \qquad (5.1)$$

where n is an integer. In practice, where there is a fixed distance between the conductors, the fields adjust themselves so that eqn. (5.1) is satisfied. The two parallel conducting surfaces from a parallel-plate waveguide and the wave is propagated between them. The direction of propagation is along the plane of the surfaces which is not the same as the direction of the ray of the plane wave. It is seen from Figure 5.3 that the spacing between points of equal phase on the wave in the direction of propagation is not the same as the characteristic wavelength of the wave. It is larger and is called the *waveguide wavelength*, denoted by λ_g.

From the diagram λ_g is given by

$$\lambda_g = \frac{\lambda_0}{\cos\theta} \qquad (5.2)$$

From Eqn. (5.1) it is seen that if

$$b < \tfrac{1}{2} n\lambda_0$$

$\sin\theta$ needs to be greater than unity and propagation cannot occur in that mode in that waveguide. At the limiting condition,

$$b = \tfrac{1}{2} n\lambda_0 \qquad (5.3)$$

$\theta = 90°$ and the waveguide wavelength is infinite. For smaller values of b, the waveguide wavelength is a mathematically imaginary quantity which can be shown by applying eqn. (4.61) to this situation. The limiting condition is called *cut-off*. Because it is always assumed that there is a fixed size of waveguide, the limiting condition is given by the frequency of the wave. The limiting frequency is called the cut-off frequency, f_c, and the corresponding characteristic wavelength is called the cut-off wavelegth, λ_c. From eqn. (5.3) the cut-off wavelength is given by

$$\lambda_c = \frac{2b}{n} \qquad (5.4)$$

and from eqn. (4.31), the corresponding cut-off frequency is given by

$$f_c = \frac{cn}{2b} \qquad (5.5)$$

Combining eqns. (5.1) and (5.2) to eliminate θ gives

$$\left(\frac{\lambda_0}{\lambda_g}\right)^2 = 1 - \left(\frac{n\lambda_0}{2b}\right)^2$$

and substituting from eqn. (5.4) gives

$$\left(\frac{\lambda_0}{\lambda_g}\right)^2 = 1 - \left(\frac{\lambda_0}{\lambda_c}\right)^2$$

or

$$\frac{1}{\lambda_g^2} = \frac{1}{\lambda_0^2} - \frac{1}{\lambda_c^2} \tag{5.6}$$

From eqn. (5.6) it is seen that the waveguide wavelength is a function of the characteristic wavelength of the electromagnetic wave and the cut-off wavelength of the waveguide. The cut-off wavelength is a function of the size and, as will be seen in Chapter 6, the shape of the waveguide. In terms of frequencies, eqn. (5.6) becomes

$$\lambda_g = \frac{c}{\sqrt{(f^2 - f_c^2)}} \tag{5.7}$$

The cut-off wavelength is also a function of the integer n, which is determined by the pattern of the fields of the electromagnetic wave propagating in the waveguide and is constant for any particular mode of propagation in the waveguide. A plot of the waveguide wavelength as a function of frequency is given in Figure 5.4. It is seen that as the frequency of the wave

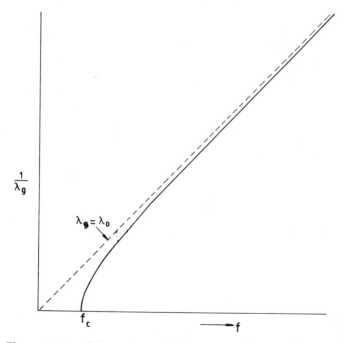

Figure 5.4 The variation of the reciprocal of the waveguide wavelength with change of frequency. Shown dotted is the characteristic wavelength for the same frequency.

increases above the cut-off frequency, so the waveguide wavelength becomes asymptotic to the characteristic wavelength and the angle θ of the ray in Figure 5.3 becomes smaller and smaller. However, as the frequency increases, so more and more additional modes are able to propagate in the waveguide.

Consider the modes that are able to propagate in any particular size of parallel plate waveguide. At low frequencies only the plane-wave type of mode is able to propagate in the waveguide. This mode is called the transmission-line mode and can propagate at all frequencies right down to zero frequency. At frequencies greater than that of the lowest cut-off frequency given by $n = 1$ in epn.(5.5), that is at frequencies greater than that given by

$$f = \frac{c}{2b}$$

the first of the waveguide modes is able to propagate as well as the transmission-line mode. When $n = 2$, at frequencies greater than that given by

$$f = \frac{c}{b}$$

two waveguide modes are able to propagate. As the frequency is increased, more and more modes are able to propagate in the waveguide.

Transmission-line modes, having the same wavelength as a plane wave and having their electromagnetic field vectors in the plane perpendicular to the direction of propagation like a plane wave, exist on any two-conductor or multi-conductor transmission line. They cannot exist in any single-conductor transmission line such as a hollow metal pipe. However, waveguide modes do exist in a single conductor transmission line. Their wavelength, the waveguide wavelength, is different from that of a plane wave at the same frequency propagating through an infinite medium of the same material filling the space in or round the waveguide. Also the electromagnetic fields do not lie solely in the plane perpendicular to the direction of propagation. Waveguide modes need to have a component of the field in the direction of propagation. However, they may often be separated into those modes with the electric field solely in the plane perpendicular to the direction of propagation, the so-called *transverse electric* or TE-modes, and those with a solely transverse magnetic field, the *transverse magnetic* or TM-modes. Using the same notation the plane wave and the transmission-line modes have *transverse electric and magnetic* fields so that these are TEM-modes.

To return to the plane-wave description of waveguide propagation in parallel-plate waveguide as shown in Figure 5.3, if the orientation of the polarisation of the plane wave is such that the magnetic field is parallel to the boundary surfaces, similar to the situation shown in Figure 4.3, the magnetic field vector always acts in a direction perpendicular to the plane of

Figure 5.3 and perpendicular to the direction of propagation. It is a transverse magnetic wave or a TM-mode in the waveguide. Similarly, if the plane wave is orientated with the electric field parallel to the boundary surfaces, there is a transverse electric wave or a TE-mode in the waveguide. Any arbitrary orientation of the waves may be described by a suitable combination of TM and TE-modes.

In parallel plate waveguide, the cut-off frequencies of the TM and TE-modes are the same. This is not true for all shapes of waveguide. The integer n in eqns. (5.4) and (5.5) gives the number of the mode in the waveguide and it is added as a subscript to the mode description. Therefore the complete list of possible modes in parallel plate waveguide is TEM-mode, TE_n-modes and TM_n-modes.

SUMMARY

Parallel-plate waveguide supports a transmission-line mode similar to a plane wave which propagates at any frequency. It can also support waveguide modes which cannot propagate below their cut-off frequency. The characteristic wavelength corresponding to the cut-off frequency is called the cut-off wavelength,

$$\lambda_c = \frac{2b}{n} \qquad (5.4)$$

The cut-off frequency is

$$f_c = \frac{cn}{2b} \qquad (5.5)$$

The waveguide wavelength is different from the characteristic wavelength:

$$\frac{1}{\lambda_g^2} = \frac{1}{\lambda_0^2} - \frac{1}{\lambda_c^2} \qquad (5.6)$$

$$\lambda_g = \frac{c}{\sqrt{(f^2 - f_c^2)}} \qquad (5.7)$$

The transmission-line mode has all the components of its fields in the transverse plane; it is a TEM-mode. The waveguide modes are TE_n-modes or TM_n-modes.

Example 5.1 An electromagnetic wave at 10 GHz propagates between two plane parallel copper sheets 4.0 cm apart. Calculate how many modes are able to propagate between the sheets and the waveguide wavelength for each mode.

Answer The characteristic wavelength is given by eqn. (4.31). Therefore

$$\lambda_0 = \frac{c}{f} = \frac{3 \times 10^8}{1 \times 10^{10}} = 3.0 \text{ cm}$$

which is also the waveguide wavelength of the TEM-mode.

142 WAVEGUIDE EFFECTS

The cut-off wavelength of the waveguide modes is given by eqn. (5.4):

$$n = 1, \quad \lambda_c = 8.0 \text{ cm}$$
$$n = 2, \quad \lambda_c = 4.0 \text{ cm}$$
$$n = 3, \quad \lambda_c = 2.67 \text{ cm}$$

As for $n = 3$ the cut-off wavelength is less than the characteristic wavelength, modes for $n > 2$ cannot propagate in this waveguide. The modes that are able to propagate in this waveguide are TEM-mode, TE_1-mode, TM_1-mode, TE_2-mode and TM_2-mode, five modes in all. Rearrangement of eqn. (5.6) gives the waveguide wavelength

$$\lambda_g = \frac{\lambda_0}{\sqrt{[1 - (\lambda_0/\lambda_c)^2]}}$$

Inserting numbers gives, when $n = 1$, $\lambda_g = 3.24$ cm, when $n = 2$, $\lambda_g = 4.54$ cm. The full list of modes and their waveguide wavelength is given in Table 5.1.

Table 5.1

Mode	n	λ_g(cm)
TEM	0	3.0
TE_1 and TM_1	1	3.24
TE_2 and TM_2	2	4.54

PROBLEM

5.1 Calculate the number of modes supported by a parallel-plate waveguide, 22.0 cm between plates, at a signal frequency of 3.0 GHz. Calculate the waveguide wavelength of each mode.

[9; 10.0, 10.27, 11.23, 13.67, 24.0 cm]

5.2 WAVE VELOCITIES

The velocity of an electromagnetic wave is the speed of an observer remaining at a point of constant phase on the wave. Therefore from eqn. (4.21)

$$\omega t - \beta z = \text{constant}$$

The velocity is given by differentiating this equation. It is called the *phase velocity* because it is the speed of a point of constant phase. It is given by

$$v_p = \frac{dz}{dt} = \frac{\omega}{\beta} \quad (5.8)$$

which is the same as eqns. (3.45) and (4.20). However, the waveguide wavelength is longer than the characteristic wavelength so that a point of constant phase in the waveguide mode of propagation travels further in the same time than a similar point on a plane wave. Therefore the phase velocity in waveguide is faster than the phase velocity of a plane wave. Since the phase velocity of a plane wave is the same as the speed of light, the phase velocity in waveguide is faster than the speed of light and the theory of relativity may appear to have been contradicted. However, phase velocity is only an apparent velocity. It is the speed of movement of a point of constant phase in the centre of a continuous wave. To carry information on the wave, the wave has to be modulated and it will be shown that the modulation does not travel at the phase velocity.

To return to the pictorial description of waveguide transmission in Figure 5.3, the plane wave is travelling along the direction of the ray at the speed of light, c. Therefore for the waveguide mode,

$$v_p = \frac{c}{\cos \theta}$$

or by geometry,

$$v_p = \frac{c \lambda_g}{\lambda_0} \tag{5.9}$$

If the phase constant of a propagating wave is related to the waveguide wavelength by a relationship similar to eqn. (4.29),

$$\lambda_g = \frac{2\pi}{\beta} \tag{5.10}$$

Substituting from eqns. (4.30) and (5.10) into eqn. (5.9) gives

$$v_p = \frac{\omega}{\beta} \tag{5.8}$$

hence confirming our earlier result. As might be expected, the phase velocity derived from the pictorial description of propagation in parallel-plate waveguide is the same as that derived from a theoretical consideration of the properties of the propagating wave.

In order to carry information on the wave, it is necessary to modulate the wave. If the wave is pulse modulated, the initial point of any disturbance travels at the speed of light but the main part of the disturbance travels at a slower speed called the group velocity, denoted by v_g. Consider an amplitude-modulated wave where the modulating frequency is very small compared to the carrier frequency. Let the carrier frequency be denoted by ω and the modulating frequency by $\delta\omega$. Then the modulated wave consists of a carrier wave and two sidebands, having the angular frequencies, ω,

$\omega + \delta\omega$ and $\omega - \delta\omega$. The corresponding phase constants are chosen to be β, $\beta + \delta\beta$ and $\beta - \delta\beta$. Then the combined propagating wave is

$$E = E_0 \cos(\omega t - \beta z) + \tfrac{1}{2} E_0 m \cos(\omega t + \delta\omega t - \beta z - \delta\beta z)$$
$$+ \tfrac{1}{2} E_0 m \cos(\omega t - \delta\omega t - \beta z + \delta\beta z)$$

Therefore

$$E = E_0 [1 + m \cos(\delta\omega t - \delta\beta z)] \cos(\omega t - \beta z) \qquad (5.11)$$

From eqn. (5.11) it is seen that the carrier wave is still propagating at the phase velocity given by eqn. (5.8) but the modulating wave is propagating at another velocity given by

$$v_g = \frac{\delta\omega}{\delta\beta} \qquad (5.12)$$

However, β is a function of ω and not vice versa, so that in the limit of small increments, eqn. (5.12) becomes

$$v_g = \frac{1}{\partial\beta/\partial\omega} \qquad (5.13)$$

To return to the pictorial description of waveguide propagation in Figure 5.3, information travels along the ray path. Therefore, by geometry, the group velocity is given by

$$v_g = c \cos \theta$$

which, by substitution from eqn. (5.2), gives

$$v_g = c \frac{\lambda_0}{\lambda_g} \qquad (5.14)$$

Substituting from eqns. (4.30) and (5.10) into eqn. (5.14) gives

$$v_g = c^2 \frac{\beta}{\omega} \qquad (5.15)$$

We now relate eqn. (5.13) to eqn. (5.15). Substituting the frequencies and from eqn. (5.10) into eqn. (5.6) gives

$$\beta^2 = \frac{\omega^2}{c^2} - \frac{\omega_c^2}{c^2}$$

Therefore

$$\frac{\partial \beta}{\partial \omega} = \frac{\omega}{c^2 \beta}$$

which when substituted into eqn. (5.13) gives eqn. (5.15). This shows, as would be expected, that the expression for the group velocity derived mathematically is the same as that derived from the pictorial description of waveguide propagation.

Nothing actually moves at the phase velocity so that relativity is not violated. Consider waves on the surface of a volume of water. The waves appear to be travelling away from the disturbance that causes them but the water is not moving in the direction of propagation of the wave. The water is ocillating vertically whereas the waves are travelling horizontally across the surface of the water. Therefore the phase velocity of the water waves can be much greater than the velocity of any movement of the water. For waveguide transmission, the power flows along the ray path at the group velocity and nothing actually moves at the phase velocity.

For waveguide propagation, in general,

$$v_p > c > v_g$$

From eqns. (5.9) and (5.14)

$$v_p v_g = c^2 \tag{5.16}$$

For a plane wave propagating in free space and for the two-conductor and multi-conductor transmission-line mode where $\lambda_g = \lambda_0$,

$$v_p = v_g = c$$

SUMMARY

Phase velocity is the velocity of a point of constant phase on the wave:

$$v_p = \frac{\omega}{\beta} = c \frac{\lambda_g}{\lambda_0} \tag{5.8) (5.9}$$

The group velocity is the velocity of any modulation or power flow on the wave:

$$v_g = \frac{1}{\partial \beta / \partial \omega} = c \frac{\lambda_0}{\lambda_g} \tag{5.12) (5.14}$$

$$\lambda_g = \frac{2\pi}{\beta} \tag{5.10}$$

$$v_p v_g = c^2 \tag{5.16}$$

146 WAVEGUIDE EFFECTS

Example 5.2 Calculate the phase velocities and group velocities of the various modes that can propagate between two plane parallel copper sheets 4.0 cm apart at 10 GHz.

Answer The problem is the same as that posed in Example 5.1. The waveguide wavelength and the characteristic wavelength were calculated in the answer to that example. The wavelengths are substituted into eqns. (5.9) and (5.14). Repeating and expanding Table 5.1, the velocities are as shown in Table 5.2.

Table 5.2

Mode	n	λ_g (cm)	v_p ($\times 10^8$ m/s)	v_g ($\times 10^8$ m/s)
TEM	0	3.0	3.0	3.0
TE_1 and TM_1	1	3.24	3.24	2.78
TE_2 and TM_2	2	4.54	4.54	1.98

PROBLEMS

5.2 Calculate the wave velocities of the modes that propagate in the parallel-plate waveguide of problem 5.1 at 3.0 GHz.

[3.0, 3.0; 3.08, 2.92; 3.37, 2.67; 4.10, 2.19; 7.20, 1.25 $\times 10^8$ m/s]

5.3 A waveguide 6.0 cm long is measured to have a phase length of 180° at 3.0 GHz A modulated carrier wave at 3.0 GHz is sent along the waveguide. Calculate the time taken for the modulation on the wave to traverse the length of waveguide.

[2.4 $\times 10^{-10}$ s]

5.3 TRANSMISSION LINE FIELDS

For low-frequency use, the simplest transmission line is the two-wire line which is shown together with its electric and magnetic fields in Figure 5.5. The field patterns are more easily obtained from the d.c. flux and potential lines. The lines of constant potential are obtained from the d.c. potential difference between the conductors. They are perpendicular to the lines of electric flux density so that these equipotential lines are parallel to the magnetic field intensity lines of the electromagnetic field. Similarly, the d.c. lines of constant flux are parallel to the electric flux density vector at each point in the field and are perpendicular to the equipotential lines so that these lines are parallel to the electric field intensity lines of the electromagnetic field. Hence, the picture of the d. c. flux and potential lines also gives a

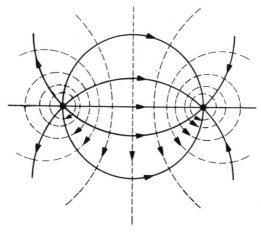

Figure 5.5 Two-wire transmission line: ⎯⎯→ electric field; ----→ magnetic field (see p. ix).

picture of the electric and magnetic fields due to an electromagnetic wave on the same line. The field lines in Figure 5.5 are circles and segments of circles which are easily drawn by geometric construction but there is no simple mathematical expression for the components of the electromagnetic field at any point. For high-frequency use, the two-wire line has the disadvantage that the fields extend to infinity and power is lost by radiation from the line.

Probably the commonest completely enclosed transmission line is the coaxial cable. If a parallel-plate transmission line is bent, the electric field remains perpendicular to the plane of the conductors and the magnetic field remains parallel to the conductors as shown in Figure 5.6. Continuing the bending to complete the circle gives the fields in a coaxial cable, as shown in Figure 5.7. From a consideration of static electromagnetic fields, the electric field intensity and the magentic field intensity are derived from the d.c. potential difference and electric current:

$$E = \frac{V}{r \ln(a/b)} \qquad (5.17)$$

$$H = \frac{I}{2\pi r} \qquad (5.18)$$

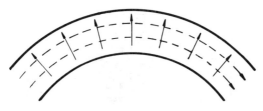

Figure 5.6 Bent parallel plate waveguide: ⎯⎯→ electric field; ----→ magnetic field (see p. ix).

Figure 5.7 Coaxial transmission line: ———→ electric field; – – – –→ magnetic field (see p. ix).

where a and b are the outer and inner radii of the coaxial line respectively. At d.c., V and I are controlled by the conditions of the circuit external to the coaxial line and eqns. (5.17) and (5.18) are independent. At a.c. the electromagnetic fields are related by Maxwell's equations. Assuming that the field in the transmission line is that of a TEM-mode, there are no longitudinal components of either field and eqn. (4.27) still applies. Therefore E and H in any transmission line are related by

$$E = \eta H \tag{5.19}$$

Substitution from eqns. (5.17)–(5.19) show that the characteristic impedance of a coaxial line is given by

$$Z_0 = \frac{V}{I} = \frac{\eta}{2\pi} \ln \frac{a}{b} \tag{5.20}$$

The result in eqn. (5.20) may also be obtained by calculating the capacitance and inductance per unit length and using eqn. (2.13).

If the coaxial line shown in Figure 5.7 is modified to the rectangular shape shown in Figure 5.8, the fields take the shape shown. These fields are very similar to those of the line consisting of a narrow strip conductor between two parallel earthed conducting planes as shown in Figure 5.9. This transmission line is called *stripline*. It is often constructed by a process of etching the conductor shape on one side of a copper-coated board similar to a

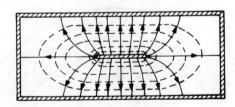

Figure 5.8 Rectangular coaxial transmission line. ———→ electric field; – – – –→ magnetic field.

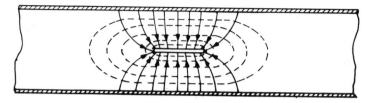

Figure 5.9 The electromagnetic fields in stripline: ———→ electric field; ----→ magnetic field.

printed circuit. Another copper-coated board is then mounted over the top to provide the screening. Stripline is useful as a high-frequency transmission line into which circuit components may be inserted. The field strength decays very rapidly away from the edge of the conducting strip and there is very little interaction between conductors that are more than the dielectric thickness apart. The stripline shown in Figure 5.9 is sometimes called triplate line because of its three conductors. It has the advantage that the electromagnetic field is totally enclosed and there is no radiation. However, it is difficult to find means to insert components into triplate line circuits or to undertake circuit adjustments.

The unsymmetrical stripline called *microstrip* is shown in Figure 5.10. It is easier to construct than triplate line and it is easier to design the circuit physically to accommodate circuit components. In a microminiature form, it is used with unencapsulated semiconductor devices to make microwave oscillator or amplifier circuits. In microwave integrated circuits, the interconnections within the chip have to be designed as microstrip lines. Microstrip has the disadvantage that part of the field is in the air space above the dielectric so that the line ought to be enclosed in a metal box. Because the fields of the wave are partly in dielectric and partly in air, the wave is not a pure TEM transmission line mode and the fields are difficult to calculate. However, except at the very highest microwave frequencies, it is satisfactory to assume that a quasi-TEM mode is propagating in the microstrip. An

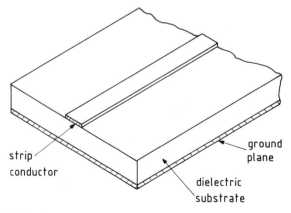

Figure 5.10 Microstrip line.

approximate form of the fields in microstrip is shown in Figure 5.11. Even with microstrip line, there is difficulty in making circuit connections to earth because the backing conductor is the earth conductor. There is another transmission line which has a particular simplicity in adding components in shunt across the line which is called *slotline*, which is shown in Figure 5.12.

For all these two-conductor transmission lines the TEM transmission line mode is the mode that propagates down to the lowest frequencies. However, waveguide modes are able to propagate at higher frequencies. The lines are most frequently used when the TEM-mode is the only mode that can propagate in the line and in design it is assumed that this is so. Then it is necessary to ensure that the higher-order waveguide modes are not able to propagate. In microstrip and similar lines it is sufficient to assume that the line approximates to a parallel-plate waveguide and to use eqn. (5.5) for the cut-off frequencies of the waveguide modes. For microstrip it may also be necessary to treat any enclosing box as a section of rectangular waveguide whose cut-off frequencies are given in the next chapter, in eqn. (6.27). The calculation of the cut-off frequencies of the waveguide modes in coaxial line is very complicated but the cut-off wavelength of the lowest

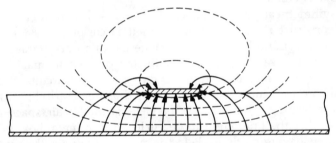

Figure 5.11 Showing the fields in microstrip line: ⟶ electric field; ----→ magnetic field (see p. ix).

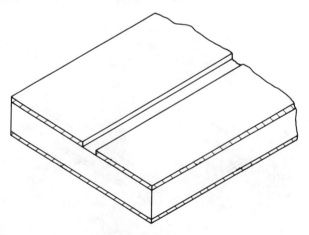

Figure 5.12 Slotline with a ground plane.

order waveguide mode is given approximately by

$$\lambda_c \approx \pi(a + b) \qquad (5.21)$$

SUMMARY

The commonest transmission lines which support the TEM-mode are:

two-wire line
coaxial line
triplate stripline
microstrip
slotline.

Example 5.3 A triplate stripline consists of a stripline conductor of negligible thickness sandwiched between two PTFE slabs, each 2.0 mm thick. The earth conductors are on the outside of the PTFE slabs. The relative permittivity of PTFE is 2.0. Calculate the maximum frequency at which this stripline might be expected to operate only in the transmission line mode.

Answer The cut-off frequency for the waveguide modes in parallel plate waveguide is given by eqn. (5.5). For the lowest-order mode, $n = 1$. A waveguide mode could propagate between the two earth conductors in the space where there is no central stripline. Therefore the dimension in eqn. (5.5), $b = 4.0$ mm, and the cut-off frequency is given by

$$f_c = \frac{c}{2b\sqrt{\varepsilon_r}} = \frac{3 \times 10^8}{2 \times 4 \times 10^{-3}\sqrt{2}} = 26.5 \text{ GHz}$$

PROBLEMS

5.4 A microstrip line is constructed on an alumina substrate, $\varepsilon_r = 10$, 1.0 mm thick. Calculate the maximum frequency at which this microstrip line might be expected to operate only in the transmission line mode.

[47 GHz]

5.5 A coaxial line has a characteristic impedance of 50 Ω. Its insulating material has the properties of air with a relative permittivity of one. The inner diameter of the outer conductor is 3.0 mm. Calculate the cut-off frequency of the first waveguide mode of propagation on the line.

[46 GHz]

5.4 WAVEGUIDE

If there is only a single conductor, the transmission line mode cannot propagate but, for many shapes of conductor, waveguide modes are able to propagate. Probably the simplest waveguide is the hollow metal pipe. Waveguide modes are able to propagate inside any shape of hollow metal pipe. Provided that the circumference of the pipe is a continuous conductor, the electromagnetic wave is totally enclosed and there can be no radiation from the hollow metal waveguide. Geometrically, the simplest waveguide is the circular metal tube as shown in Figure 5.13. It is analysed mathematically in Chapter 6 to determine the cut-off frequencies of the waveguide modes and the field patterns of the modes. The electric and magnetic fields in the cross-section of the waveguide for the simplest mode are shown in Figure 5.13. However, there is one major disadvantage of circular waveguide. There is nothing to stop the direction of polarisation of the fields in the waveguide from changing. The field is launched into the waveguide with a particular direction of polarisation. However, any slight perturbations in the regular shape of the waveguide can cause the direction of polarisation to rotate so that the direction of polarisation at the output from the waveguide may not be the same as that at the input. There is also only a small frequency range over which the lowest-order mode is the only mode that can propagate. This is important since it is usual to operate waveguide in that condition where the lowest order mode is the only mode operating.

The commonest shape of hollow metal pipe to be used as waveguide is rectangular with a 2:1 ratio of sides, as shown in Figure 5.14. It is analysed in detail in Chapter 6. There is a 2:1 ratio of frequencies over which the lowest-order mode is the only mode able to propagate in rectangular

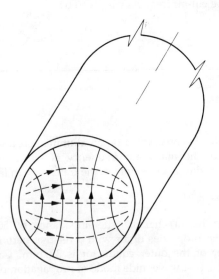

Figure 5.13 Circular waveguide showing the electric and magnetic field lines for the simplest mode of propagation in the waveguide: ⎯⎯⎯→ electric field; ⎯ ⎯ ⎯→ magnetic field.

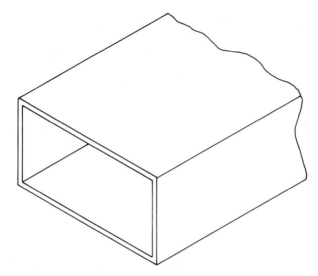

Figure 5.14 Rectangular waveguide.

waveguide. The field pattern in the waveguide is also governed by the shape of the waveguide so that the polarisation of the wave is confined by the waveguide shape. These two properties make rectangular the most useful shape for waveguide. A variation on the rectangular waveguide is the ridge waveguide as shown in Figure 5.15. It has a very much greater frequency range over which only the lowest-order mode is able to propagate. It is used when electromagnetic waves with a very wide frequency range need to be propagated along a single waveguide pipe and coaxial line has an unacceptably high attenuation. In the double ridge waveguide, most of the power is carried in the narrow space between the ridges where the fields approximate to those of a plane wave.

A variation on the circular waveguide is the elliptical waveguide. For each mode equivalent to a mode in circular waveguide, there are two different modes in the elliptical waveguide called the even and odd modes. The field patterns appropriate to the even and odd modes of the lowest-order mode in elliptical waveguide are shown in Figure 5.16. Provided that the ellipticity is large, there is little conversion between the two modes shown in Figure 5.16 because their cut-off frequencies and waveguide wavelengths are different.

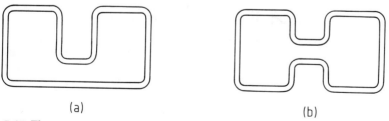

Figure 5.15 The cross-sectional shape of ridge waveguide: (a) single ridge; (b) double ridge.

154 WAVEGUIDE EFFECTS

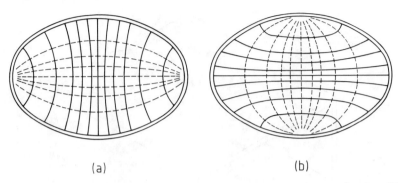

Figure 5.16 The field patterns appropriate to the lowest-order modes in elliptical waveguide: (a) even mode, (b) odd mode. ⟶ electric field; ---→ magnetic field (see p. ix).

A further variation of the rectangular waveguide is to enclose a slotline in a rectangular waveguide. It is called the finline and is shown in Figure 5.17. It has many of the properties of the ridge waveguide and at the same time it provides a simple mechanism for mounting active components in a waveguide. It is used at the higher microwave frequencies where the attenuation of microstrip is too high.

There is a mode that propagates parallel to a plane surface, whether it is the conducting surface shown in Figure 5.2 or the plane boundary between two dielectric media shown in Figure 4.3. It is called the surface mode and has an exponential decay of the fields in the direction perpendicular to the surface. In the dielectric slab waveguide having no metal conducting surfaces, waveguide propagation occurs in the space between two parallel plane surfaces in the same way that propagation occurs in parallel-plate

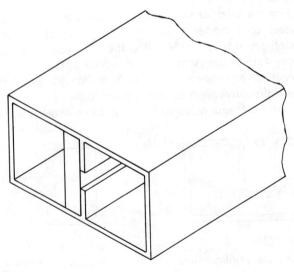

Figure 5.17 Finline.

waveguide. The dielectric slab waveguide is analysed in Chapter 7. In the direction perpendicular to the plane of the slab, there is a sinusoidal variation of field strength within the slab and an exponential decay of the fields outside the slab. An infinite number of waveguide modes are possible similar to the parallel-plate waveguide.

Waveguide modes can propagate within any shape of dielectric rod. The relative permittivity between the rod and its surroundings determines the relative concentration of the electromagnetic fields in the two media. However, the waveguide modes have a periodic variation of the fields in the transverse plane inside the rod and an exponential decay of the fields outside. Common shapes of dielectric rod waveguide are circular rod and rectangular rod and simple variations. The optical fibre waveguide as discussed in Chapter 7 is a variation of the circular dielectric rod waveguide. The image waveguide is a rectangular dielectric rod waveguide supported on a plane conducting surface. It is used for circuits operating above 50 GHz.

Other shapes of solid metal waveguides have been used when a limited radiation field is required. Such a waveguide is useful to provide a communication channel to a vehicle running on a track. Two such waveguides are the H-guide and the Y-guide where the waveguide is a continuous metal bar of H or Y cross-section. The electromagnetic fields are strongly confined within the overall profile of the waveguide yet there is space for a probe on the vehicle to couple to the fields on the waveguide. Three simple shapes of waveguide that are in common use and which are relatively easy to analyse theoretically are discussed in detail in Chapters 6 and 7.

SUMMARY

The common shapes of hollow metal waveguide are

circular
rectangular
the ridge waveguide is a rectangular waveguide with ridges
the finline is a rectangular waveguide with slotline

Waveguide modes are able to propagate along a plane metallic surface or along the surface between two dielectric media, such as

circular dielectric rod waveguide
rectangular dielectric rod waveguide
optical fibre waveguide
image guide.

CHAPTER 6

Hollow Metal Waveguide

Aims: to give a theoretical derivation of the fields and properties of the propagating modes in rectangular and circular waveguide and their applications.

6.1 RECTANGULAR WAVEGUIDE

The rectangular waveguide is the simplest shape of waveguide to analyse theoretically and one in which the theoretical results give a good description of reality. Because a hollow metal waveguide is used to enclose the electromagnetic radiation, only the inside shape and size of the waveguide is relevant to the theoretical derivation. A rectangular waveguide of size a by b is shown in Figure 6.1 related to a system of rectangular coordinates. The wave equation is obtained by the solution of Maxwell's equations in rectangular coordinates for a non-conducting medium with no stored charges. The wave equations are given by eqns. (4.12) and (4.13), which are rewritten here:

$$\nabla^2 \mathbf{E} + \omega^2 \mu \varepsilon \mathbf{E} = 0 \qquad (6.1)$$

$$\nabla^2 \mathbf{H} + \omega^2 \mu \varepsilon \mathbf{H} = 0 \qquad (6.2)$$

The plane wave has no components of the fields in the direction of propagation; however, in Section 5.1, it has been shown that waveguide modes need to have a component of the field in the direction of propagation. These are called the *longitudinal* component of the field which, in the orientation of the waveguide shown in Figure 6.1, are the z-directed components of the field. There is an equation for each of the three components of the electric field and another three equations for the components of the magnetic field. Provided ε and μ are constants, eqns. (6.1) and (6.2) show that each component of **D** and **B** also satisfies a similar wave equation. There are twelve similar wave equations, each one defining a different component of one of the electromagnetic fields. It is possible to seek a solution of any one of these equations, but having determined an expression for one component of one of the field quantities, all the other components of the field quantities will be related to it by Maxwell's curl equations, eqns. (4.10) and (4.11), or by the material properties, eqns. (4.5) and (4.6). Experience has shown that the mathematics is easiest to manipulate if solutions to eqns. (6.1) and (6.2) are obtained in terms of longitudinal components of the electric and magnetic fields.

The z-component of eqn. (6.1) written in its expanded form is similar to eqn. (4.15):

$$\frac{\partial^2 E_z}{\partial x^2} + \frac{\partial^2 E_z}{\partial y^2} + \frac{\partial^2 E_z}{\partial z^2} = -\omega^2 \mu \varepsilon E_z \tag{6.3}$$

and a similar equation for H_z is

$$\frac{\partial^2 H_z}{\partial x^2} + \frac{\partial^2 H_z}{\partial y^2} + \frac{\partial^2 H_z}{\partial z^2} = -\omega^2 \mu \varepsilon H_z. \tag{6.4}$$

These equations are solved by the separation of variables technique. E_z is a function of all three space coordinates. Let it be a function of x multiplied by a function of y multiplied by a function of z. Let these functions be $f_1(x)$, $f_2(y)$ and $f_3(z)$ respectively; then

$$E_z = f_1(x) f_2(y) f_3(z) \tag{6.5}$$

Differentiating eqn. (6.5) gives

$$\frac{\partial^2 E_z}{\partial x^2} = f_1''(x) f_2(y) f_3(z)$$

where the prime denotes differentiation with respect to the argument. There are two similar equations for y and z. Substituting these relationships into eqn. (6.3) and dividing through by E_z from eqn. (6.5) gives

$$\frac{f_1''(x)}{f_1(x)} + \frac{f_2''(y)}{f_2(y)} + \frac{f_3''(z)}{f_3(z)} = -\omega^2 \mu \varepsilon = -k^2, \quad \text{say} \tag{6.6}$$

It is seen that each term on the left-hand side of eqn. (6.6) is a function of only one of the space coordinates and yet their sum is the constant we have called k^2. As eqn. (6.6) is true for any combination of x, y and z representing any point in space inside the waveguide, each term in eqn. (6.6) must itself be equal to a constant. Therefore let,

$$\frac{f_1''(x)}{f_1(x)} = k_x^2; \quad \frac{f_2''(y)}{f_2(y)} = k_y^2; \quad \frac{f_3''(z)}{f_3(z)} = k_z^2 \tag{6.7}$$

where k_x, k_y and k_z are constants related by substituting into eqn. (6.6):

$$k_x^2 + k_y^2 + k_z^2 = -k^2 \tag{6.8}$$

HOLLOW METAL WAVEGUIDE

The z-direction is the direction of propagation so a distinction is made between k_z and the other two constants, whose combined effect is given by another constant k_c. Therefore let

$$k_x^2 + k_y^2 = -k_c^2 \tag{6.9}$$

The subscript c is used because the constant k_c defines the cut-off conditions in the waveguide. Substitution into eqn. (6.8) gives

$$k_z^2 = k_c^2 - k^2 \tag{6.10}$$

The z-dependent part of eqn. (6.7) may be written

$$\frac{\partial^2 E_z}{\partial z^2} = k_z^2 E_z$$

This equation has the same form as eqns. (3.4), (3.5) and (4.16) and its solution is similar to eqn. (4.21),

$$E_z = E_0 \exp[j(\omega t - \beta z)] \tag{6.11}$$

where

$$\beta = jk_z = \pm j\sqrt{(k_c^2 - k^2)} = \pm\sqrt{(k^2 - k_c^2)} \tag{6.12}$$

If β is real, eqn. (6.11) is a description of the field of a wave propagating in the waveguide. As far as variation with respect to time and the direction of propagation is concerned, the wave has the same properties as a plane wave or as a transmission line wave. If β is imaginary, however, it is the equation of a wave having an exponential decay of field strength in the z-direction. There is no propagation along the waveguide. Lossless propagation cannot occur and the constant k_z behaves as if it were an attenuation constant. There is no sinusoidal variation of fields in the propagation direction, just an exponential decay. The transition between the two defines the cut-off condition, i.e. $\beta = 0$. Then

$$k_c = k = \omega\sqrt{(\mu\varepsilon)} \tag{6.13}$$

The cut-off frequency is given by

$$f_c = \frac{k_c}{2\pi\sqrt{(\mu\varepsilon)}} \tag{6.14}$$

and the cut-off wavelength by

$$\lambda_c = \frac{2\pi}{k_c} \tag{6.15}$$

From eqn. (4.29), the plane wave wavelength is given by

$$\lambda = \frac{2\pi}{\omega\sqrt{(\mu\varepsilon)}} = \frac{2\pi}{k} \tag{6.16}$$

and the waveguide wavelength by eqn. (5.10):

$$\lambda_g = \frac{2\pi}{\beta} = \frac{2\pi}{jk_z} \tag{6.17}$$

Substituting from eqns. (6.15)–(6.17) into eqn. (6.10) gives

$$\frac{1}{\lambda_c^2} + \frac{1}{\lambda_g^2} = \frac{1}{\lambda^2} \tag{6.18}$$

If the waveguide is filled with air, it is a very good approximation to a vacuum with constants $\varepsilon = \varepsilon_0$ and $\mu = \mu_0$, then $\lambda = \lambda_0$, and eqn. (6.18) becomes the same as eqn. (5.6). Equation (6.18) may be rearranged to give an expression for the waveguide wavelength in terms of the cut-off wavelength and the characteristic wavelength:

$$\lambda_g = \frac{\lambda_0}{\sqrt{[1 - (\lambda_0/\lambda_c)^2]}} \tag{6.19}$$

The solution to eqn. (6.3) takes the form

$$E_z = f_1(x) f_2(y) \exp[j(\omega t - \beta z)] \tag{6.20}$$

In the x- and y-directions, it is necessary to find a solution to the differential equation which satisfies the boundary conditions given by the waveguide walls. For the ideal waveguide, the boundary to the wave is two pairs of perfectly conducting walls. The rectangular pipe is shown in Figure 6.1 and the electric field intensity parallel to the walls is zero at the walls. Therefore E_z is zero at the walls. In eqn. (6.20), the x- and y-dependence of the fields takes the same form as the z-dependence. One possible solution to eqn. (6.3) has an exponential variation of the fields in the x- and y-directions similar to that given in eqn. (4.17). However, that solution does not satisfy the boundary conditions of the waveguide walls. There is an alternative solution which has a periodic variation of the fields in the x- and y-directions. The solution is a linear combination of trigonometric functions; therefore let

$$f_1(x) = A \sin(jk_x x) + B \cos(jk_x x) \tag{6.21}$$

$$f_2(y) = C \sin(jk_y y) + D \cos(jk_y y) \tag{6.22}$$

where A, B, C and D are arbitrary constants. The boundary is a perfectly

160 HOLLOW METAL WAVEGUIDE

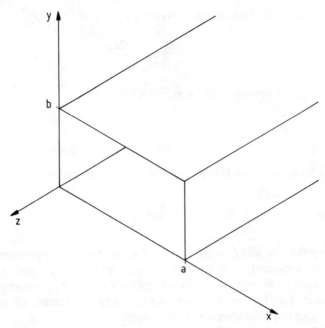

Figure 6.1 Rectangular waveguide, cross-sectional dimensions a by b, showing its relationship to the axes of the rectangular coordinates.

conducting wall at $x = 0$ and $x = a$. Therefore eqn. (6.21) must be zero at $x = 0$ and $x = a$. The first condition gives $B = 0$ and the second

$$A \sin(jk_x a) = 0$$

Therefore

$$jk_x = \frac{m\pi}{a} \quad \text{or} \quad k_x = -\frac{jm\pi}{a} \tag{6.23}$$

where m is an integer. Similarly, eqn. (6.21) must be zero at $y = 0$ and $y = b$. Therefore $D = 0$ and

$$jk_y = \frac{n\pi}{b} \quad \text{or} \quad k_y = -\frac{jn\pi}{b} \tag{6.24}$$

where n is an integer. Combining the two amplitude constants into E_0 and substituting into eqn. (6.20) gives the full solution to eqn. (6.3) incorporating the boundary conditions for the rectangular waveguide shown in Figure 6.1,

$$E_z = E_0 \sin\frac{m\pi x}{a} \sin\frac{n\pi y}{b} \exp[j(\omega t - \beta z)] \tag{6.25}$$

The cut-off conditions for this mode of propagation in rectangular

RECTANGULAR WAVEGUIDE

waveguide are obtained by substituting for k_x and k_y from eqns. (6.23) and (6.24) into eqn. (6.9):

$$k_c^2 = \left(\frac{m\pi}{a}\right)^2 + \left(\frac{n\pi}{b}\right)^2$$

Therefore

$$\lambda_c = \frac{1}{\sqrt{[(m/2a)^2 + (n/2b)^2]}} \tag{6.26}$$

and

$$f_c = c\sqrt{[(m/2a)^2 + (n/2b)^2]} \tag{6.27}$$

where m and n are integers which may take any value. Each different combination of m and n constitutes a separate solution to eqn. (6.3) and a separate solution to Maxwell's equations for rectangular waveguide. In mathematical terms, each solution to the differential equation is called an eigenfunction and m and n are the eigennumbers. Each solution gives rise to a different mode of propagation in the waveguide and m and n are the mode numbers. E_z gives rise to a family of modes having a longitudinal component of the electric field and no longitudinal component of the magnetic field so that they are transverse magnetic or TM-modes. Therefore eqn. (6.25) describes one component of the electric field of the TM_{mn}-mode. There is also another family of modes, with a longitudinal magnetic field and no longitudinal electric field, obtained by finding a solution to eqn. (6.4); some of the other components of the fields of the wave are needed before the boundary condition equations for that family of modes can be solved. These are the TE-modes whose propagating conditions will be determined after expressions for the other components of the fields are obtained.

SUMMARY

Rectangular waveguide has a perfectly conducting boundary of rectangular shape which contains the electromagnetic wave. Maxwell's equations reduce to the wave equations

$$\frac{\partial^2 E_z}{\partial x^2} + \frac{\partial^2 E_z}{\partial y^2} + \frac{\partial^2 E_z}{\partial z^2} = -\omega^2 \mu \varepsilon E_z = -k^2 E_z \tag{6.3}$$

$$\frac{\partial^2 H_z}{\partial x^2} + \frac{\partial^2 H_z}{\partial y^2} + \frac{\partial^2 H_z}{\partial z^2} = -\omega^2 \mu \varepsilon E_z = -k^2 E_z \tag{6.4}$$

$$\beta = \pm\sqrt{(k^2 - k_c^2)} \tag{6.12}$$

162   HOLLOW METAL WAVEGUIDE

The cut-off condition is

$$k_c = \omega\sqrt{(\mu\varepsilon)} \qquad (6.13)$$

$$\frac{1}{\lambda_c^2} + \frac{1}{\lambda_g^2} = \frac{1}{\lambda^2} \qquad (6.18)$$

Waveguide wavelength

$$\lambda_g = \frac{\lambda_0}{\sqrt{[1 - (\lambda_0/\lambda_c)^2]}} \qquad (6.19)$$

Rectangular waveguide boundary conditions are such that, for a waveguide of dimensions $a \times b$,

$$\lambda_c = \frac{1}{\sqrt{[(m/2a)^2 + (n/2b)^2]}} \qquad (6.26)$$

$$f_c = c\sqrt{[(m/2a)^2 + (n/2b)^2]} \qquad (6.27)$$

The longitudinal component of the electric field intensity of the TM_{mn}-mode is

$$E_z = E_0 \sin\frac{m\pi x}{a} \sin\frac{n\pi y}{b} \exp[j(\omega t - \beta z)] \qquad (6.25)$$

Example 6.1 Calculate the cut-off frequency of the TM_{11}-mode in rectangular waveguide having the inside dimensions 2.0 cm by 1.0 cm.

Answer The solution is obtained by substitution into eqn. (6.27) for $m = 1$ and $n = 1$. Therefore

$$f_c = 3.0 \times 10^8 \times 10^2 \sqrt{(\tfrac{1}{16} + \tfrac{1}{4})} = 10^{10} \times 3\sqrt{5}/4 = 16.8 \text{ GHz}$$

PROBLEM

6.1 Calculate the cut-off frequencies of the TM_{11}, TM_{12}, TM_{31} and TM_{22}-modes in rectangular waveguides of sizes, all given in inches, 9.75 by 4.875, 4.30 by 2.15, 0.90 by 0.40 and 0.28 by 0.14. These are all standard sizes of drawn copper waveguide.

[1.36, 2.49, 2.18, 2.70; 3.07, 5.66, 4.95, 6.14; 16.2, 30.2, 24.6, 32.3; 47.2, 87.0, 76.0, 94.3 GHz]

6.2 FIELD COMPONENTS IN RECTANGULAR WAVEGUIDE

In order to find the relationships between the different components of the electromagnetic wave in the waveguide and in particular to relate all the other field components to the longitudinal components of the field, it is necessary to return to Maxwell's curl equations, eqns. (4.10) and (4.11), separated into their component form. The variation of the fields in the direction of propagation, the z-direction, is already known, so that the relationship may be entered into the equations,

$$\frac{\partial}{\partial z} = -j\beta$$

The curl equations separated into their component form are given in eqns. (4.22) and (4.23). When the known z-variation is entered, these equations become

$$\frac{\partial E_z}{\partial y} + j\beta E_y = -j\omega\mu H_x \qquad (6.28)$$

$$-j\beta E_x - \frac{\partial E_z}{\partial x} = -j\omega\mu H_y \qquad (6.29)$$

$$\frac{\partial E_y}{\partial x} - \frac{\partial E_x}{\partial y} = -j\omega\mu H_z \qquad (6.30)$$

$$\frac{\partial H_z}{\partial y} + j\beta H_y = j\omega\varepsilon E_x \qquad (6.31)$$

$$-j\beta H_x - \frac{\partial H_z}{\partial x} = j\omega\varepsilon E_y \qquad (6.32)$$

$$\frac{\partial H_y}{\partial x} - \frac{\partial H_x}{\partial y} = j\omega\varepsilon E_z \qquad (6.33)$$

Two pairs of simultaneous equations in E_y and H_x and in E_x and H_y are formed by eqns. (6.28) and (6.32) and (6.29) and (6.31) respectively. Therefore they may easily be solved to obtain expressions for the other components of the fields in terms of the longitudinal components. Rewriting the above equations in pairs and simplifying gives

$$\begin{cases} \beta E_y + \omega\mu H_x = j\dfrac{\partial E_z}{\partial y} & (6.28a) \\[6pt] \omega\varepsilon E_y + \beta H_x = j\dfrac{\partial H_z}{\partial x} & (6.32a) \end{cases}$$

$$\begin{cases} \beta E_x - \omega\mu H_y = j\dfrac{\partial E_z}{\partial x} & (6.29a) \\[6pt] \omega\varepsilon E_x - \beta H_y = -j\dfrac{\partial H_z}{\partial y} & (6.31a) \end{cases}$$

From eqn. (6.12),

$$\omega^2\mu\varepsilon - \beta^2 = k_c^2$$

Therefore the solutions of the simultaneous equations are

$$E_y = \frac{j}{k_c^2}\left(-\beta\frac{\partial E_z}{\partial y} + \omega\mu\frac{\partial H_z}{\partial x}\right) \quad (6.34)$$

$$H_x = \frac{j}{k_c^2}\left(\omega\varepsilon\frac{\partial E_z}{\partial y} - \beta\frac{\partial H_z}{\partial x}\right) \quad (6.35)$$

$$E_x = \frac{-j}{k_c^2}\left(\beta\frac{\partial E_z}{\partial x} + \omega\mu\frac{\partial H_z}{\partial y}\right) \quad (6.36)$$

$$H_y = \frac{-j}{k_c^2}\left(\omega\varepsilon\frac{\partial E_z}{\partial x} + \beta\frac{\partial H_z}{\partial y}\right) \quad (6.37)$$

It is seen that the longitudinal components of the fields appear in eqns. (6.34)–(6.37) as independent variables, and Maxwell's equations do not provide any other connection between them. Therefore either E_z or H_z may be put equal to zero without affecting the other, giving two independent sets of field components. The wave for which E_z exists and $H_z = 0$ is a TM-mode and that for which H_z exists and $E_z = 0$ is a TE-mode. Substitution of $H_z = 0$ and E_z from eqn. (6.25) into eqns. (6.34)–(6.37) gives the field components for the TM_{mn}-modes in rectangular waveguides:

$$\begin{aligned}
E_x &= -\frac{j\beta E_0}{k_c^2}\frac{m\pi}{a}\cos\frac{m\pi x}{a}\sin\frac{n\pi y}{b}\exp[j(\omega t - \beta z)] \\
E_y &= -\frac{j\beta E_0}{k_c^2}\frac{n\pi}{b}\sin\frac{m\pi x}{a}\cos\frac{n\pi y}{b}\exp[j(\omega t - \beta z)] \\
E_z &= E_0\sin\frac{m\pi x}{a}\sin\frac{n\pi y}{b}\exp[j(\omega t - \beta z)] \\
H_x &= \frac{j\omega\varepsilon E_0}{k_c^2}\frac{n\pi}{b}\sin\frac{m\pi x}{a}\cos\frac{n\pi y}{b}\exp[j(\omega t - \beta z)] \\
H_y &= -\frac{j\omega\varepsilon E_0}{k_c^2}\frac{m\pi}{a}\cos\frac{m\pi x}{a}\sin\frac{n\pi y}{b}\exp[j(\omega t - \beta z)] \\
H_z &= 0
\end{aligned} \quad (6.38)$$

In order to find the components of the fields for the TE-modes, it is necessary to find a solution to eqn. (6.4) that satisfies the boundary conditions. The general solution is similar to eqn. (6.20):

$$H_z = f_1(x)f_2(y)\exp[j(\omega t - \beta z)] \quad (6.39)$$

FIELD COMPONENTS IN RECTANGULAR WAVEGUIDE

where $f_1(x)$ and $f_2(y)$ are given by eqns. (6.21) and (6.22). The boundary conditions are that the tangential component of the electric field intensity is zero at the walls. Therefore E_y is zero at $x = 0$ and $x = a$ and E_x is zero at $y = 0$ and $y = b$. For the TE-modes, E_y is given by eqn. (6.34) with E_z identically equal to zero. Therefore at $x = 0$ and $x = a$,

$$\frac{\partial H_z}{\partial x} = 0 \tag{6.40}$$

Similarly from eqn. (6.36), at $y = 0$ and $y = b$,

$$\frac{\partial H_z}{\partial y} = 0 \tag{6.41}$$

Applying the boundary condition, eqn. (6.40), to eqn. (6.21) shows that $A = 0$ and eqn. (6.23) still applies. Similarly, applying eqn. (6.41) to eqn. (6.22) shows that $C = 0$ and eqn. (6.24) still applies. Combining the two amplitude constants into H_0, and substituting into eqn. (6.39), gives the full solution to eqn. (6.4) incorporating the boundary conditions for the rectangular waveguide shown in Figure 6.1:

$$H_z = H_0 \cos\frac{m\pi x}{a} \cos\frac{n\pi y}{b} \exp[j(\omega t - \beta z)] \tag{6.42}$$

Substitution of $E_z = 0$ and H_z from eqn. (6.42) into eqns. (6.34)–(6.37) gives the field components for the TE$_{mn}$-mode in rectangular waveguide:

$$\left.\begin{aligned}
E_x &= \frac{j\omega\mu H_0}{k_c^2} \frac{n\pi}{b} \cos\frac{m\pi x}{a} \sin\frac{n\pi y}{b} \exp[j(\omega t - \beta z)] \\
E_y &= -\frac{j\omega\mu H_0}{k_c^2} \frac{m\pi}{a} \sin\frac{m\pi x}{a} \cos\frac{n\pi y}{b} \exp[j(\omega t - \beta z)] \\
E_z &= 0 \\
\\
H_x &= \frac{j\beta H_0}{k_c^2} \frac{m\pi}{a} \sin\frac{m\pi x}{a} \cos\frac{n\pi y}{b} \exp[j(\omega t - \beta z)] \\
H_y &= \frac{j\beta H_0}{k_c^2} \frac{n\pi}{b} \cos\frac{m\pi x}{a} \sin\frac{n\pi y}{b} \exp[j(\omega t - \beta z)] \\
H_z &= H_0 \cos\frac{m\pi x}{a} \cos\frac{n\pi y}{b} \exp[j(\omega t - \beta z)]
\end{aligned}\right\} \tag{6.43}$$

The cut-off conditions for the TE$_{mn}$-modes are the same as those for the TM$_{mn}$-modes because eqns. (6.23) and (6.24) apply to both sets of modes. Therefore the cut-off wavelength and the cut-off frequency are given by eqns. (6.26) and (6.27) respectively for both sets of modes. The lowest mode numbers for the TM-modes are $m = 1$ and $n = 1$ because if either m or n were zero E_z would be zero and there would be no electromagnetic wave in the waveguide. However, for the TE-modes, either m or n may be zero, so that the TE_{10}-mode is the lowest-order mode in rectangular waveguide. If it is assumed that the rectangular waveguide has an aspect ratio of 2:1, i.e. $a = 2b$, then for the TE_{10}-mode, $\lambda_c = 2a$, for the TE_{01}-mode and the TE_{20}-mode, $\lambda_c = a$ and for the TE_{11}-mode and the TM_{11}-mode, $\lambda_c = 2a/\sqrt{5}$. There is an octave over which the lowest-order mode is the only mode that is able to propagate in the waveguide. This lowest-order mode is called the *dominant mode*. For any two-conductor or multi-conductor transmission line, the TEM-mode is the dominant mode which propagates down to the lowest frequencies. Then the waveguide modes are the higher-order modes. In rectangular waveguide, the TE_{10}-mode is the dominant mode and there is a 2:1 frequency range over which it is the only mode which can propagate in the waveguide. Rectangular waveguide is useful because it has this wide frequency band over which the dominant mode is able to propagate alone. There is a large range of standard sizes of rectangular waveguide to provide a size suitable for operation for any frequency between 300 MHz and 300 GHz. The properties of the dominant mode are discussed in the next section.

SUMMARY

The field components of the TM_{mn}-mode are given in eqn. (6.38).
The longitudinal component of the magnetic field of the TE_{mn}-mode is

$$H_z = H_0 \cos\frac{m\pi x}{a} \cos\frac{n\pi y}{b} \exp[j(\omega t - \beta z)] \qquad (6.42)$$

The field components of the TE_{mn}-mode are given in eqn. (6.43).
The TE_{10}-mode is the dominant mode in rectangular waveguide. It has the cut-off wavelength $\lambda_c = 2a$.

Example 6.2 Calculate the frequency range over which the dominant mode is the only mode propagating in a rectangular waveguide of size 2 cm by 1 cm.

Answer Inserting the numbers $m = 1$ and $n = 0$ and $m = 2$ and $n = 0$ into eqn. (6.27) gives the required frequency range:

$$f_c = \frac{c}{2a} = \frac{3.0 \times 10^8}{4.0 \times 10^{-2}} = 7.5 \text{ GHz}$$

or

$$f_c = \frac{c}{a} = \frac{3.0 \times 10^8}{2.0 \times 10^{-2}} = 15.0 \text{ GHz}.$$

PROBLEM

6.2 Calculate the frequency range over which the dominant mode is the only mode propagating for the standard sizes of waveguide given in Problem 6.1.

[0.606, 1.21; 1.37, 2.75; 6.56, 13.12; 21.1, 42.2 GHz]

6.3 DOMINANT MODE IN RECTANGULAR WAVEGUIDE

As most waveguide systems are designed so that only the dominant mode is able to propagate, the properties of the dominant mode will be discussed in more detail. The dominant mode in rectangular waveguide is the TE_{10}-mode. Expressions for the components of the fields are obtained by substituting $m = 1$ and $n = 0$ into eqn. (6.43). Therefore

$$\left.\begin{aligned} E_x &= 0 \\ E_y &= -j\frac{\omega\mu a}{\pi} H_0 \sin\frac{\pi x}{a} \exp[j(\omega t - \beta z)] \\ E_z &= 0 \\ H_x &= j\frac{\beta a}{\pi} H_0 \sin\frac{\pi x}{a} \exp[j(\omega t - \beta z)] \\ H_y &= 0 \\ H_z &= H_0 \cos\frac{\pi x}{a} \exp[j(\omega t - \beta z)] \end{aligned}\right\} \quad (6.44)$$

It is noticed that, as well as being the dominant mode, this is also a mode with a very simple field pattern which is illustrated in Figure 6.2. There is no variation of the field strength in the y-direction. Only the y-directed component of the electric field exists and has a simple sinusoidal variation of strength across the width of the waveguide. The magnetic field exists in closed loops in the x–z plane and the whole field pattern moves along the waveguide at the phase velocity,

$$v_p = \frac{\omega}{\beta} = f\lambda_g$$

The electric field lines terminate in electric charges in the walls of the waveguide. As the wave travels along the waveguide, the currents in the waveguide walls redistribute these charges so that the electric field is always correctly terminated.

168 HOLLOW METAL WAVEGUIDE

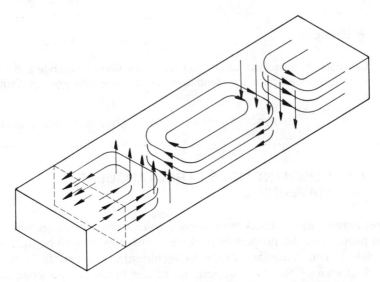

Figure 6.2 The dominant mode. The field pattern of the TE_{10}-mode in rectangular waveguide.

We now determine expressions for the currents in the waveguide walls. For simplicity, the analysis is confined to the wall currents of the dominant mode but those for other modes may be determined similarly. So far, in this chapter, it has been assumed that the waveguide walls consist of a perfect conductor so that any currents occurring in the walls are infinitely large and confined to an infinitely thin surface layer. In practice the conductivity of most waveguide metals is large but not infinite, so that the waveguide wall currents are finite and confined to the surface of the waveguide metal. For most waveguide metals at microwave frequencies, the skin depth is very small and the waveguide wall current is confined to a thin skin adjacent to the surface. As shown by eqn. (4.78), the total surface current density I_s is equal to the magnetic field strength in the air immediately adjacent to the surface. Example 4.6 shows that the ratio of the plane wave wavelength in air to the plane wave wavelength in copper is 6700:1 at 10 GHz. Therefore we are completely justified in making the assumption of plane wave propagation into the surface of the copper, so that the total surface current density in the copper of the waveguide wall is equal to the magnetic field strength in the waveguide immediately adjacent to the waveguide wall. Both the current density and the magnetic field intensity are in the plane of the waveguide wall surface and are perpendicular to one another. Therefore the waveguide wall current streamlines shown in Figure 6.3 can be drawn. It is seen that the node in the current streamlines coincides with the minimum of the electric field strength in the waveguide. The current is redistributing the charge in order to support the electric field intensity one quarter of a cycle later. However, at its minimum, the electric field strength has a maximum rate of change so that the electric circuit of the conduction current is completed by the displacement current, $\partial \mathbf{D}/\partial t$, of the electric field.

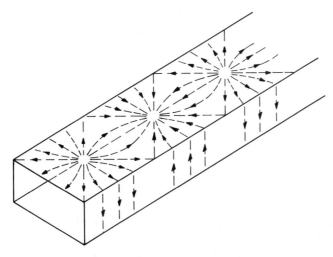

Figure 6.3 The waveguide wall currents for the TE_{10}-mode in rectangular waveguide.

There is a large range of standard sizes of rectangular waveguide for use from 300 MHz to 300 GHz. The recommended frequency of operation covers a range of 1.5:1 although the dominant mode is the sole mode of propagation over a frequency range of 2:1. At low frequencies, with the operating frequency near to the cut-off frequency, the waveguide wavelength is very long and the waveguide attenuation can become too large. Therefore the waveguide is not operated at frequencies close to the cut-off frequency. Simple radius bends and regular twists operate satisfactorily in rectangular waveguide. The bend or twist maintains the rectangular waveguide cross-section. In so far as the bent or twisted section of waveguide does not have exactly the same characteristic impedance as the straight waveguide, the bend or twist is made one half-wavelength long so that the impedance change at one end cancels the effect of the impedance change at the other end. Flexible waveguide is also available which can be bent into radius bends or twists.

Demountable lengths of waveguide are connected together by means of couplings or flanges. These are flat metal plates designed to provide the best electrical continuity for the inner surface of the waveguide. However, in regular use, the mating surfaces become damaged and tarnished and electrical continuity cannot be guaranteed, which leads to a deterioration in performance and the possibility of sparking in high-power systems. The use of a choke coupling eliminates the possibility of sparking. A diagram of a choke–plain coupling pair is shown in Figure 6.4. The choke ditch is a circular groove in the metal of the coupling which is a quarter-wavelength deep. The choke ditch is situated a quarter wavelength from the inside surface of the centre of the broad face of the waveguide. Inside the ditch radius, it is arranged that there is a distinct gap between the two lengths of waveguide. Then the short circuit at the bottom of the choke ditch is reflected to be an open circuit at the face of the coupling which in turn is

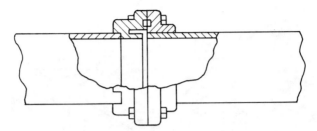

Figure 6.4 A part section through a waveguide joint using a choke–plain coupling combination (see p. ix). (Reproduced with permission from Baden Fuller, *Microwaves* (3rd edn), © 1990, Pergamon Press Ltd.)

reflected to be a short circuit at the inside surface of the waveguide. The usual choke–plain coupling only introduces a reflection coefficient of about 0.005. For high-precision measurements, it may be possible to achieve performances better than this with a combination of two plain couplings provided that they are in good condition. Otherwise the choke–plain combination gives a better performance.

In order to make measurements inside waveguide it is necessary to ensure that any slot or hole in the waveguide wall does not cut any current streamlines for the dominant mode in the waveguide. A longitudinal slot along the centreline of the broad face of the waveguide is parallel to the current streamlines at that point and no electromagnetic power is radiated out from the slot. A waveguide with such a slot may be used to measure the VSWR in the waveguide. It is called a slotted line and it is fitted with a probe to measure the electric field strength on the centreline of the waveguide, which is able to traverse along the waveguide. Such a probe is shown in Figure 6.5.

If the probe is replaced by a metal post projecting into the waveguide, it behaves as a capacitance, the value of capacitance depending on the insertion of the post. Such a post may be used for stub matching in waveguide. It provides a variable-position–variable-capacitance matching element. Alternatively the same effect may be produced by three posts at fixed positions in the waveguide. Very often the posts are screws inserted through tapped

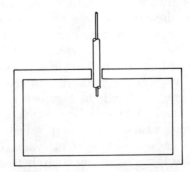

Figure 6.5 An electric probe in a slot in the centreline of the broad face of rectangular waveguide, as used in a slotted-line measuring section to measure VSWR.

holes in the waveguide wall and give rise to a device colloquially called a screw-matching section.

One example of a control device in waveguide is the variable-position vane attenuator. One way to control the electromagnetic power in waveguide is to absorb it. If a resistive vane is placed in the waveguide parallel to the narrow wall of the waveguide, it is also parallel to the electric field vector in the waveguide. The electric field generates a current in the resistive vane which is dissipated as heat. If the vane is moved from a position adjacent to the waveguide wall where the electric field strength is a minimum, to a position at the centre of the waveguide where the electric field strength is a maximum, the attenuation varies from a minimum to a maximum. One method of mounting an attenuating vane in a rectangular waveguide is shown in Figure 6.6.

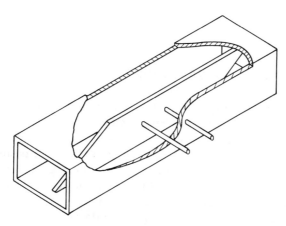

Figure 6.6 A microwave vane attenuator showing the rods for moving the vane (see p. ix). (Reproduced with permission from Baden Fuller. *Microwaves* (3rd edn), © 1990 Pergamon Press Ltd.)

SUMMARY

The TE_{10}-mode is the dominant mode in rectangular waveguide. Expressions for the field components are given in eqn. (6.44). The fields are illustrated in Figure 6.2. The electric field is perpendicular to the broad face of the waveguide. The magnetic field is in closed loops in the plane of the broad face of the waveguide.

The surface electric current density in the inside walls of the waveguide is perpendicular to and equal to the magnetic field intensity adjacent to the wall. The waveguide wall currents for the dominant mode are shown in Figure 6.3.

Example 6.3 List the positions for non-radiating slots in the walls for the dominant mode in rectangular waveguide.

Answer The current streamlines are shown in Figure 6.3. A non-radiating slot in the waveguide walls has to be parallel to the current streamlines. The only two possible positions are (i) a longitudinal slot along the centreline of the broad face of the waveguide, (ii) a vertical slot in the narrow face of the waveguide.

PROBLEMS

6.3 A special low-loss mode for propagation in rectangular waveguide is the TE_{01}-mode. Give the positions of slots in the waveguide wall which will radiate for all possible modes except the TE_{01}-mode.

[Transverse slots in the broad face of the waveguide]

6.4 CIRCULAR WAVEGUIDE

Circular pipe appears to be the simplest shape for the transmission of electromagnetic energy using waveguide modes of propagation. However, if the mode is linearly polarised, there is nothing in the waveguide to maintain the direction of polarisation. Over a long distance imperfections in the waveguide shape can rotate the direction of linear polarisation. Therefore rectangular waveguide is a preferred transmission medium for general use. However, there are situations where circular waveguide is necessary so that we discuss the properties of circular waveguide modes in this and the next section. The waveguide is shown in Figure 6.7 with its axis coincident with the axis of a system of cylindrical polar coordinates, r, θ and z. As with rectangular waveguide, the z-direction is taken as the direction of propagation. The propagation space inside the waveguide is a non-conducting medium with no stored charges so that the wave equations are still given by eqns. (6.1) and (6.2). In cylindrical polar coordinates, the Laplacian of the vector may be resolved into an equation in terms of the Laplacian of the longitudinal component of the vector and an equation involving the transverse components of the vector. There is no simple separation of the transverse components of the vector as occurs with a rectangular coordinate system. The z-component of eqn. (6.1) written in its expanded form in cylindrical polar coordinates is

$$\frac{\partial^2 E_z}{\partial r^2} + \frac{1}{r}\frac{\partial E_z}{\partial r} + \frac{1}{r^2}\frac{\partial^2 E_z}{\partial \theta^2} + \frac{\partial^2 E_z}{\partial z^2} = -\omega^2 \mu \varepsilon E_z \tag{6.45}$$

and a similar equation for H_z is

$$\frac{\partial^2 H_z}{\partial r^2} + \frac{1}{r}\frac{\partial H_z}{\partial r} + \frac{1}{r^2}\frac{\partial^2 H_z}{\partial \theta^2} + \frac{\partial^2 H_z}{\partial z^2} = -\omega^2 \mu \varepsilon H_z \tag{6.46}$$

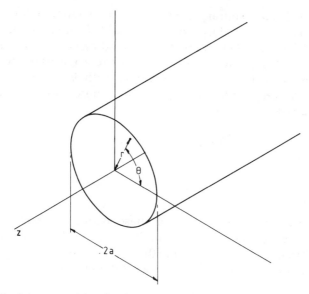

Figure 6.7 Circular waveguide of radius a showing its relationship to the axes of the cylindrical polar coordinates.

The separation of variables technique is applied to eqn. (6.45). Exponential solutions are satisfactory for both the z-variation and the θ-variation. The boundary condition in θ is that the field pattern repeats itself when $\theta = 2\pi$. Then a suitable solution is

$$E_z = f(r) \exp[j(\omega t - n\theta - \beta z)] \qquad (6.47)$$

where n is an integer. Then

$$\frac{\partial^2}{\partial \theta^2} = -n^2 \quad \text{and} \quad \frac{\partial^2}{\partial z^2} = -\beta^2$$

Therefore eqn. (6.45) becomes

$$\frac{\partial^2 E_z}{\partial r^2} + \frac{1}{r}\frac{\partial E_z}{\partial r} - \frac{n^2}{r^2}E_z - \beta^2 E_z = -k^2 E_z \qquad (6.48)$$

Substituting from eqn. (6.12) into eqn. (6.48) gives

$$\frac{\partial^2 E_z}{\partial r^2} + \frac{1}{r}\frac{\partial E_z}{\partial r} + \left(k_c^2 - \frac{n^2}{r^2}\right)E_z = 0 \qquad (6.49)$$

There is no analytic solution to eqn. (6.49) but it is in a form that occurs frequently in the analysis of scientific problems using cylindrical polar coordinates. Its solutions have been tabulated. It is one form of what is called Bessel's equation. The solution to eqn. (6.49) is

$$E_z = AJ_n(k_c r) + BY_n(k_c r) \qquad (6.50)$$

where A and B are arbitrary constants of integration whose values are determined by the boundary conditions and J_n and Y_n are the *Bessel functions* of order n of the first and second kind respectively. n is one of the mode numbers of the waveguide modes in circular waveguide. As we are mainly interested in the lower-order modes with $n = 0$ or 1, $J_0(x)$, $Y_0(x)$, $J_1(x)$ and $Y_1(x)$ are plotted in Figure 6.8. These Bessel functions have distinct properties when the argument is zero:

$$J_0(0) = 1$$
$$J_n(0) = 0, \quad n > 0$$
$$Y_n(0) = -\infty, \quad n \geq 0.$$

For larger arguments, the values of these Bessel functions oscillate like decaying sine waves and, apart from $Y_n(x)$ near to zero argument, their value is less than one. Equation (6.50) is a solution to eqn. (6.49) which, when combined with eqn. (6.47), gives a solution to eqn. (6.45):

$$E_z = [AJ_n(k_c r) + BY_n(k_c r)] \exp[j(\omega t - n\theta - \beta z)] \quad (6.51)$$

The value of k_c is governed by the dimensions of the waveguide and determines the cut-off frequency for that waveguide.

Reference to Figure 6.7 shows that the boundary condition is a perfectly conducting waveguide wall at $r = a$. Then the component of the electric field parallel to the wall is zero at the wall. Therefore E_θ and E_z are both zero at $r = a$. For the TM-modes having a longitudinal component of the electric field, $E_z = 0$ at $r = a$. Therefore, from eqn. (6.51),

$$AJ_n(k_c a) + BY_n(k_c a) = 0 \quad (6.52)$$

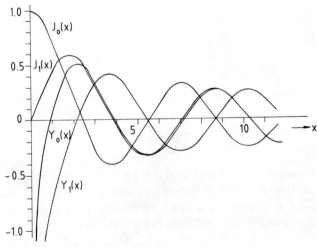

Figure 6.8 Bessel functions of the first and second kinds of order 0 and 1 (see p. ix).

The electric field must be continuous and finite at the centre of the waveguide so that no term in $Y_n(x)$ can be allowed to exist in eqn. (6.51). Therefore $B = 0$ and the other boundary condition is

$$J_n(k_c a) = 0 \tag{6.53}$$

Then the longitudinal component of the electric field of the TM$_n$-modes is

$$E_z = E_0 J_n(k_c r) \exp[j(\omega t - n\theta - \beta z)] \tag{6.54}$$

where k_c is determined by the solution to eqn. (6.53). As can be seen from Figure 6.8, there are an infinite number of zeros of $J_n(x)$. These zeros are numbered starting with 1 for the zero having the smallest value of argument. This provides the second number m in the mode numbers for the modes in circular waveguide, so that eqn. (6.54) is one component of the field of the TM$_{nm}$-mode in circular waveguide. The integer n is the order of the Bessel function in eqn. (6.54) and is a measure of the circumferential variation of the field pattern. The integer m is the number of the zero of the Bessel function of order n and is a measure of the radial variation of the field pattern. It is necessary to determine an expression for E_θ before the boundary conditions for the TE-modes can be determined.

As with rectangular waveguide, Maxwell's curl equations, eqns. (4.10) and (4.11), are used to determine the relationships between the different components of the fields. Writing these in their component form in cylindrical polar coordinates, eqn. (4.10) becomes

$$\left. \begin{aligned} \frac{1}{r} \frac{\partial E_z}{\partial \theta} - \frac{\partial E_\theta}{\partial z} &= -j\omega\mu H_r \\ \frac{\partial E_r}{\partial z} - \frac{\partial E_z}{\partial r} &= -j\omega\mu H_\theta \\ \frac{1}{r} E_\theta + \frac{\partial E_\theta}{\partial r} - \frac{1}{r} \frac{\partial E_r}{\partial \theta} &= -j\omega\mu H_z \end{aligned} \right\} \tag{6.55}$$

and eqn. (4.11) becomes

$$\left. \begin{aligned} \frac{1}{r} \frac{\partial H_z}{\partial \theta} - \frac{\partial H_\theta}{\partial z} &= j\omega\varepsilon E_r \\ \frac{\partial H_r}{\partial z} - \frac{\partial H_z}{\partial r} &= j\omega\varepsilon E_\theta \\ \frac{1}{r} H_\theta + \frac{\partial H_\theta}{\partial r} - \frac{1}{r} \frac{\partial H_r}{\partial \theta} &= j\omega\varepsilon E_z \end{aligned} \right\} \tag{6.56}$$

176   HOLLOW METAL WAVEGUIDE

In eqns. (4.10) and (4.11) the time dependence in eqn. (6.54) has already been taken into consideration. If the z and θ dependence is also incorporated into eqns. (6.55) and (6.56), they become

$$-\frac{jn}{r} E_z + j\beta E_\theta = -j\omega\mu H_r \tag{6.57}$$

$$-j\beta E_r - \frac{\partial E_z}{\partial r} = -j\omega\mu H_\theta \tag{6.58}$$

$$\frac{1}{r} E_\theta + \frac{\partial E_\theta}{\partial r} + \frac{jn}{r} E_r = -j\omega\mu H_z \tag{6.59}$$

$$-\frac{jn}{r} H_z + j\beta H_\theta = j\omega\varepsilon E_r \tag{6.60}$$

$$-j\beta H_r - \frac{\partial H_z}{\partial r} = j\omega\varepsilon E_\theta \tag{6.61}$$

$$\frac{1}{r} H_\theta + \frac{\partial H_\theta}{\partial r} + \frac{jn}{r} H_r = j\omega\varepsilon E_z \tag{6.62}$$

In a way similar to that used while working in rectangular coordinates for the fields inside rectangular waveguide, eqns. (6.57) and (6.61) and eqns. (6.58) and (6.60) form two pairs of simultaneous equations in E_θ and H_r and in E_r and H_θ respectively. Solving these equations gives

$$E_\theta = \frac{1}{k_c^2}\left(j\omega\mu\frac{\partial H_z}{\partial r} - \frac{\beta n}{r} E_z\right) \tag{6.63}$$

$$H_r = \frac{1}{k_c^2}\left(-j\beta\frac{\partial H_z}{\partial r} + \frac{\omega\varepsilon n}{r} E_z\right) \tag{6.64}$$

$$E_r = \frac{1}{k_c^2}\left(-\frac{\omega\mu n}{r} H_z - j\beta\frac{\partial E_z}{\partial r}\right) \tag{6.65}$$

$$H_\theta = \frac{1}{k_c^2}\left(-\frac{\beta n}{r} H_z - j\omega\varepsilon\frac{\partial E_z}{\partial r}\right) \tag{6.66}$$

As with rectangular waveguides, and in fact for all hollow metal waveguides, eqns. (6.63)–(6.66) show that the waveguide modes may be separated into those modes having a longitudinal component of the electric field, the TM-modes, and those having a longitudinal component of the magnetic field, the TE-modes. Substitution from eqn. (6.54) and $H_z = 0$ into eqns. (6.63)–(6.66) gives expressions for the components of the fields of the TM_{nm}-mode in circular waveguide. They are

$$\left.\begin{aligned}
E_r &= -\frac{j\beta}{k_c} E_0 J'_n(k_c r) \exp[j(\omega t - n\theta - \beta z)] \\
E_\theta &= -\frac{\beta n}{k_c^2} E_0 \frac{1}{r} J_n(k_c r) \exp[j(\omega t - n\theta - \beta z)] \\
E_z &= E_0 J_n(k_c r) \exp[j(\omega t - n\theta - \beta z)] \\
H_r &= \frac{\omega \varepsilon n}{k_c^2} E_0 \frac{1}{r} J_n(k_c r) \exp[j(\omega t - n\theta - \beta z)] \\
H_\theta &= -\frac{j\omega\varepsilon}{k_c} E_0 J'_n(k_c r) \exp[j(\omega t - n\theta - \beta z)] \\
H_z &= 0
\end{aligned}\right\} \quad (6.67)$$

where the prime indicates the differentiation of the Bessel function with respect to its argument.

For the TE-modes, there is no longitudinal component of the electric field, so that the boundary condition becomes $E_\theta = 0$ at $r = a$. From eqn. (6.63), this gives

$$\frac{\partial H_z}{\partial r} = 0$$

at $r = a$. An expression for H_z is given by a solution to eqn. (6.46). The solution is similar to eqn. (6.54):

$$H_z = H_0 J_n(k_c r) \exp[j(\omega t - n\theta - \beta z)] \quad (6.68)$$

and the boundary condition is

$$J'_n(k_c a) = 0 \quad (6.69)$$

The derivative of the Bessel function is also an oscillatory function having an infinite number of zeros. The mode nomenclature of the TE-modes is similar to that of the TM-modes in that the second subscript is the number of the zero of the function $J'_n(x)$.

The cut-off frequency of the various modes is determined from the values of the arguments of the Bessel functions when the Bessel functions are zero. The values for the first five modes in circular waveguide are given in Table 6.1. If the value of the argument is denoted by x and the waveguide radius is a, the cut-off condition is given by

$$k_c = \frac{x}{a}$$

178 HOLLOW METAL WAVEGUIDE

and by substitution into eqn. (6.15)

$$\lambda_c = \frac{2\pi a}{x}$$

Table 6.1 Zeros of Bessel functions, $J_n(x)$ and $J'_n(x)$.

Mode	mth zero	of	value of x
TE_{11}	1	$J'_1(x)$	1.84
TM_{01}	1	$J_0(x)$	2.40
TE_{21}	1	$J'_2(x)$	3.05
TM_{11}	1	$J_1(x)$	3.83
TE_{01}	1	$J'_0(x)$	3.83

Substitution from eqn. (6.68) and $E_z = 0$ into eqns. (6.63) to (6.66) gives expressions for the components of the fields of the TE_{nm}-mode in circular waveguide. They are

$$\left. \begin{aligned} E_r &= -\frac{\omega\mu n}{k_c^2} H_0 \frac{1}{r} J_n(k_c r) \exp[j(\omega t - n\theta - \beta z)] \\ E_\theta &= \frac{j\omega\mu}{k_c} H_0 J'_n(k_c r) \exp[j(\omega t - n\theta - \beta z)] \\ E_z &= 0 \\ H_r &= -\frac{j\beta}{k_c} H_0 J'_n(k_c r) \exp[j(\omega t - n\theta - \beta z)] \\ H_\theta &= -\frac{\beta n}{k_c^2} H_0 \frac{1}{r} J_n(k_c r) \exp[j(\omega t - n\theta - \beta z)] \\ H_z &= H_0 J_n(k_c r) \exp[j(\omega t - n\theta - \beta z)] \end{aligned} \right\} \quad (6.70)$$

The TE_{11}-mode is the dominant mode in circular waveguide. The field pattern in the cross-section of the waveguide is shown in Figure 6.9. The field distribution is very similar to that shown in Figure 6.2, which shows that the field pattern for the dominant modes in rectangular and circular waveguide are similar. From Table 6.1 it is seen that there is a much smaller separation between the cut-off frequency of the dominant mode and that of the next mode than in standard rectangular waveguide. This gives another reason why rectangular waveguide is the preferred shape of waveguide for general transmission purposes.

SUMMARY

Circular waveguide has a perfectly conducting circular boundary of radius a. In cylindrical polar coordinates, Maxwell's equations reduce to the wave

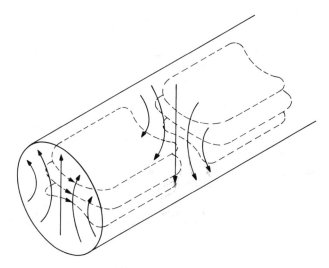

Figure 6.9 Fields of the TE_{11}-mode in circular waveguide. —— electric field; - - - - - magnetic field.

equations:

$$\frac{\partial^2 E_z}{\partial r^2} + \frac{1}{r}\frac{\partial E_z}{\partial r} + \frac{1}{r^2}\frac{\partial^2 E_z}{\partial \theta^2} + \frac{\partial^2 E_z}{\partial z^2} = -\omega^2 \mu\varepsilon E_z \quad (6.45)$$

$$\frac{\partial^2 H_z}{\partial r^2} + \frac{1}{r}\frac{\partial H_z}{\partial r} + \frac{1}{r^2}\frac{\partial^2 H_z}{\partial \theta^2} + \frac{\partial^2 H_z}{\partial z^2} = -\omega^2 \mu\varepsilon H_z \quad (6.46)$$

Separation of variables gives an exponential solution for the z and θ dimensions:

$$E_z = f(r)\exp[j(\omega t - n\theta - \beta z)] \quad (6.47)$$

and Bessel's equation,

$$\frac{\partial^2 E_z}{\partial r^2} + \frac{1}{r}\frac{\partial E_z}{\partial r} + \left(k_c^2 - \frac{n^2}{r^2}\right)E_z = 0 \quad (6.49)$$

with the solution

$$E_z = [AJ_n(k_c r) + BY_n(k_c r)]\exp[j(\omega t - n\theta - \beta z)] \quad (6.51)$$

and a similar expression for H_z.

The circular waveguide boundary conditions are given by the solution to the equations,

$$J_n(k_c a) = 0 \quad \text{for the TM-modes} \quad (6.53)$$
$$J'_n(k_c a) = 0 \quad \text{for the TE-modes} \quad (6.69)$$

The field components of the TM$_{nm}$-modes are given in eqn. (6.67). The field components of the TE$_{nm}$-modes are given in eqn. (6.70). The TE$_{11}$-mode is the dominant mode in circular waveguide. It has the cut-off wavelength $\lambda_c = 3.4a$, where a is the radius of the waveguide.

Example 6.4 Calculate the frequency range over which the dominant mode is the only mode propagating in a circular waveguide of 2 cm inside diameter.

Answer The radius of the waveguide is $a = 1$ cm. The cut-off frequency is obtained from the value of x from Table 6.1. Therefore

$$f_c = \frac{c}{\lambda_c} = \frac{cx}{2\pi a} = \frac{3 \times 10^8 \times 1.84}{2\pi \times 10^{-2}} = 8.79 \text{ GHz} \quad \text{for the TE}_{11}\text{-mode}$$

$$= \frac{3 \times 10^8 \times 2.40}{2\pi \times 10^{-2}} = 11.5 \text{ GHz} \quad \text{for the TM}_{01}\text{-mode}$$

PROBLEM

6.4 Putting $n = 0$ into eqn. (6.67), sketch the field pattern for the TM$_{01}$-mode in circular waveguide.

[See Figure 6.13]

6.5 DOMINANT MODE IN CIRCULAR WAVEGUIDE

The TE$_{11}$-mode in circular waveguide is the dominant mode. It is seen from Table 6.1 and from the answer to Example 6.4 that the dominant mode is the only mode propagating in the circular waveguide over a frequency range of 1.3:1 as opposed to the range of 2:1 for rectangular waveguide. This is one reason why circular waveguide is much less useful for general propagation purposes than rectangular waveguide. The picture in Figure 6.9 shows the field pattern in the waveguide for a linearly polarised wave at one instant of time. However, the θ dependence in eqn. (6.70) depicts a circularly polarised wave in the waveguide. The whole field pattern in Figure 5.13 is rotating in time and if time were frozen the maximum electric field traces out a helix in space as shown in Figure 6.10. However, there is also an alternative solution to the wave equation which is a sensible description for the dominant mode or to any other mode where $n = 1$. The solution replaces the exponential variation with θ by a trigonometric variation with θ. H_z has the θ dependence, $\cos \theta$. Then the components of the fields for the TE$_{11}$-

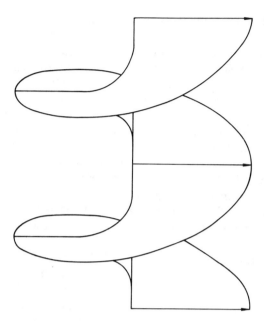

Figure 6.10 Locus of the electric field vector of a circularly polarised wave for two wavelengths (see p. ix). (Reproduced with permission from Baden Fuller, *Microwaves* (3rd edn), © 1990 Pergamon Press Ltd.)

mode in circular waveguide are given by

$$\left. \begin{aligned} E_r &= j\frac{\omega\mu}{k_c^2} H_0 \frac{1}{r} J_1(k_c r) \sin\theta \exp[j(\omega t - \beta z)] \\ E_\theta &= j\frac{\omega\mu}{k_c} H_0 J_1'(k_c r) \cos\theta \exp[j(\omega t - \beta z)] \\ E_z &= 0 \\ H_r &= -j\frac{\beta}{k_c} H_0 J_1'(k_c r) \cos\theta \exp[j(\omega t - \beta z)] \\ H_\theta &= j\frac{\beta}{k_c^2} H_0 \frac{1}{r} J_1(k_c r) \sin\theta \exp[j(\omega t - \beta z)] \\ H_z &= H_0 J_1(k_c r) \cos\theta \exp[j(\omega t - \beta z)] \end{aligned} \right\} \quad (6.71)$$

The wave described in eqn. (6.71) is a linearly polarised wave as shown in Figure 6.9. As can be seen from Figure 6.9 and Figure 5.13, the field at the centre of the waveguide approximates to a plane wave, so that for this dominant mode the plane of polarisation has the same definition as that given in Section 4.4. The angular dependence of $\exp(-jn\theta)$, where n is an integer in eqns. (6.67) and (6.70) depict fields that are rotating with time or with z-variation. These are called circularly polarised modes in the waveguide. In the wave equation, n can be a positive or a negative integer but the order of the Bessel function is always positive because the wave equation only has n^2. Signed values of n are used elsewhere in eqns. (6.67) and (6.70). Then a positive value for n gives *positive circular polarisation* and a negative value for n gives *negative circular polarisation*. However, the terms positive and negative are dependent on the particular system of coordinates used to describe the fields of the wave. The terms *right-* and *left-hand circular polarisation* are preferred. Then a right-hand circularly polarised wave has a positive n in a right-handed cylindrical coordinate system and a left-hand circularly polarised wave has a negative value for n. From a descriptive point of view for a right-hand circularly polarised wave if time is frozen, the locus of the maximum of the electric field, traces a helix in space whose rotation is counter-clockwise. Alternatively, if the observer remains in one place so that z remains constant, the maximum of the electric field rotates clockwise.

The wave described in eqn. (6.71) is called a linearly polarised wave. For the TE_{11}-mode, the plane of polarisation is the plane containing the maximum of the electric field vector. For other modes with more complicated field patterns a definition of the plane of polarisation is more difficult but linear polarisation still means that the field pattern in the waveguide does not rotate but remains in the same orientation everywhere in the waveguide. Alternatively circular polarisation describes a wave whose field pattern does rotate.

Comparison between the field patterns given in Figure 6.2 and Figure 6.9 show the great similarity between the two dominant modes in the two different shapes of waveguide. Then the dominant mode in circular waveguide may be launched from the dominant mode in rectangular waveguide. Because the mode names are different, the launcher is called a mode transformer. It transforms between the TE_{10}-mode in rectangular waveguide and the TE_{11}-mode in circular waveguide. A simple taper between the two shapes of waveguide makes a satisfactory mode transformer. Because of the possibility that imperfections in the circular waveguide can cause rotation of the plane of polarisation of the wave in the circular waveguide, a mode transformer in circular waveguide is usually accompanied by a mode absorber which consists of an absorbing vane perpendicular to the electric field vector in the circular waveguide. The absorbing vane absorbs any component of the electric field which is parallel to it while the wanted component which is perpendicular to the vane is unaffected by it. The vane is positioned so that it absorbs any perpendicularly polarised mode in the circular waveguide which could not propagate in the rectangular waveguide.

An example of the use of circular waveguide is the *rotary attenuator*. A diagram of the attenuator is shown in Figure 6.11. The rectangular to circular waveguide transformer is drawn as a taper. The associated mode absorber is essential to the rotary attenuator. The centre section of the attenuator also contains a mode absorber and it is free to rotate compared with the rest of the attenuator. When it is parallel to the other mode absorbers, they are all

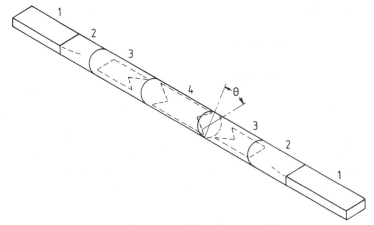

Figure 6.11 Rotary attenuator. (1) Input rectangular waveguide. (2) Rectangular to circular waveguide mode transformer. (3) Mode absorber, consisting of an absorbing vane mounted parallel to the broad wall of the rectangular waveguide. (4) Rotating section of circular waveguide also containing a mode absorber (see p. ix). (Reproduced with permission from Baden Fuller, *Microwaves* (3rd edn), © 1990 Pergamon Press Ltd.)

perpendicular to the electric field vector of the TE_{11}-mode in the circular waveguide and there is no attenuation. When the centre section makes an angle θ with the position of minimum attenuation, the linearly polarised wave in the circular waveguide is resolved into two components parallel and perpendicular to the mode absorber. The parallel component is absorbed and the perpendicular component is transmitted without loss. If the input signal has an amplitude E_0, as shown in Figure 6.12, then the transmitted signal leaving the rotating section has an amplitude $E_0 \cos \theta$. On entry to the second fixed section, the signal is resolved a second time and the final signal out is

$$E_{out} = E_0 \cos^2 \theta$$

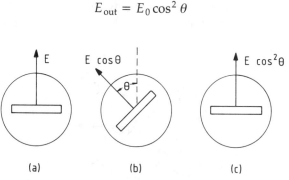

Figure 6.12 Some relative field strengths in the rotary attenuator: (a) a section through the input mode absorber; (b) a section through the centre section; (c) a section through the output mode absorber (see p. ix). (Reproduced with permission from Baden Fuller, *Microwaves* (3rd edn), © 1990 Pergamon Press Ltd.)

184   HOLLOW METAL WAVEGUIDE

If the attenuator is calibrated in decibels, the attenuation is

$$\text{attenuation} = 40 \log_{10} (\sec \theta) \quad \text{dB} \tag{6.72}$$

The rotary attenuator is a direct reading device where attenuation is directly related to an angle. Therefore it is self-calibrating provided that care is taken to eliminate sources of error.

Mode absorbers used in conjunction with the dominant mode in circular waveguide may be used to increase the frequency range of operation of the TE_{11} dominant mode in that they also selectively absorb the second mode that can propagate in the waveguide. The second mode is the TM_{01}-mode and its field components are given by putting $n = 0$ into eqn. (6.67):

$$\left.\begin{aligned}
E_r &= -\frac{j\beta}{k_c} E_0 J_0'(k_c r) \exp[j(\omega t - \beta z)] \\
E_\theta &= 0 \\
E_z &= E_0 J_0(k_c r) \exp[j(\omega t - \beta z)] \\
H_r &= 0 \\
H_\theta &= -\frac{j\omega\varepsilon}{k_c} E_0 J_0'(k_c r) \exp[j(\omega t - \beta z)] \\
H_z &= 0
\end{aligned}\right\} \tag{6.73}$$

The field pattern in the transverse section of the waveguide is shown in Figure 6.13. Because the TM_{01}-mode is circularly symmetric in the circular waveguide, it may be used to construct a rotating joint in waveguide. Many radar systems need to have a rotating aerial system and a waveguide joint is needed between the aerial and the rest of the equipment. Also the electrical phase length through the joint needs to be independent of angle. In this rotating joint there is a mode transformer from the TE_{10}-mode in rectangular waveguide to the TM_{01}-mode in circular waveguide. The joint is shown in Figure 6.14. The loops of magnetic field in the rectangular waveguide

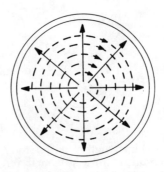

Figure 6.13 The field pattern of the TM_{01}-mode in circular waveguide. —— electric field; ----- magnetic field.

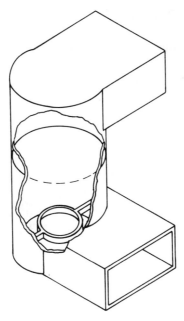

Figure 6.14 A rotating joint between two sections of rectangular waveguide making use of the TM_{01}-mode in circular waveguide.

transform into the circular magnetic field in the circular waveguide. The electric field in the rectangular waveguide transforms into the longitudinal component of the electric field in the circular waveguide. The metal ring in the circular waveguide suppresses the TE_{11}-mode and other modes in the circular waveguide. Because the TM_{01}-mode in the circular waveguide is circularly symmetric, the electric length through the rotating joint is independent of the angle of rotation of the joint.

SUMMARY

The TE_{11}-mode is the dominant mode in circular waveguide and it has a similar field pattern to the TE_{10} dominant mode in rectangular waveguide so that the two may be launched from one another.

In a circularly polarised wave, the field pattern rotates in the waveguide.

In a linearly polarised wave the field pattern maintains a constant orientation in the waveguide.

For the TE_{11}-mode, the field at the centre of the waveguide approximates to that of a plane wave. The plane of polarisation is that containing the maximum electric field vector.

A mode absorber absorbs that linearly polarised wave whose plane of polarisation lies in the plane of the absorber. The perpendicular wave is transmitted without loss.

HOLLOW METAL WAVEGUIDE

The law of a rotary attenuator is

$$\text{attenuation} = 40 \log_{10}(\sec \theta) \quad \text{dB} \quad (6.72)$$

The TM_{01}-mode is a circularly symmetric mode which may be used to construct a rotary joint.

Example 6.5 Show that a circularly polarised wave may be generated by the addition of two linearly polarised waves perpendicular to one another and that a linearly polarised wave may be generated from two circularly polarised waves of opposite hand.

Answer The circularly polarised TE_{11}-mode in circular waveguide is described by eqn. (6.70) and the linearly polarised mode by eqn. (6.71). Considering only the θ dependence of the field quantities,

$$\exp(j\theta) = \cos \theta + j \sin \theta$$

Therefore a circularly polarised wave is the sum of two equal linearly polarised waves perpendicular to one another and 90° out of phase. Similarly,

$$2 \cos \theta = \exp(j\theta) + \exp(-j\theta)$$

Therefore a linearly polarised wave is the sum of two equal amplitude circularly polarised waves of opposite hand.

PROBLEM

6.5 Calculate the attenuation of the rotary attenuator for the following angles of the rotating vane, 10°, 30°, 45°, 60° and 80°.

[0.27, 2.50, 6.02, 12.0, 30.4 dB]

6.6 RESONANT CAVITY

A closed metal box acts as a microwave resonator and is called a resonant cavity. It has all the characteristics of a microwave resonant circuit. The reader will be familiar with the concept of resonant frequency and Q-factor. At low frequencies the resonance curve is obtained from measurements of current or voltage when the other is maintained constant. At higher frequencies, when the electrical signal exists more in the form of electromagnetic waves and fields than as voltages and currents that can be measured, the resonance curve is measured by the variation of impedance or reflection coefficient in a line. A typical resonance curve is shown in Figure 6.15. The

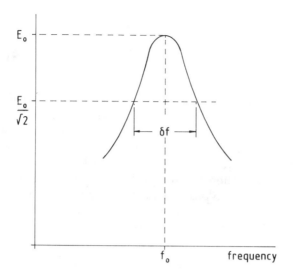

Figure 6.15 Resonance curve (see p. ix).

Q-factor of a microwave resonant cavity is usually measured from the width of the resonance curve. The formulae are

$$Q = \frac{f_0}{\delta f} = 2\pi f_0 \frac{\text{energy stored}}{\text{power dissipated}} \qquad (6.74)$$

A length of transmission line with a short circuit at each end acts as a resonant cavity. The voltage on a line terminated in a short circuit is shown in Figure 3.4. It is seen that the voltage on the line is zero at every multiple of a half-wavelength from the short circuit. From the answer to Example 3.4 or by substitution into eqn. (3.83) of the conditions for a line terminated in a short circuit, the voltage on a short-circuited line takes the form

$$V = V_{\max} \sin \beta z \qquad (6.75)$$

The second short circuit is placed at any of the zeros of the voltage on the line. Then the line is an integral number of half-wavelengths long,

$$\beta z = n\pi$$

where n is an integer. Therefore, substituting from eqn. (3.14),

$$z = \frac{n\pi}{\beta} = \frac{n\pi\lambda}{2\pi} = \frac{n\lambda}{2} \qquad (6.76)$$

This shows that any shortcircuited line has an infinite number of resonant frequencies. If v is the velocity of the wave on the line, substitution from eqn. (3.15) shows that the resonant frequencies of the short-circuited line are given by

$$f_0 = \frac{nv}{2z} \qquad (6.77)$$

At low frequencies, any resonant line will be excessively long and its losses large so that the Q-factor is low. A resonant circuit provides a better resonance with a higher Q-factor. At higher frequencies, lines also usually tend to have high losses so that resonant lengths of line have low Q-factors. At high frequencies, waveguide provides the lowest loss transmission medium so that high-frequency resonance is best provided by a short-circuited length of waveguide. A typical resonant cavity using rectangular waveguide is shown in Figure 6.16. It is seen immediately that this is a symmetrical system so that eqn. (6.8) can be written

$$\omega^2 \mu\varepsilon = k^2 = -k_x^2 - k_y^2 - k_z^2$$

and the resonant frequency of the cavity is given by

$$f_0 = \frac{1}{2\pi} \sqrt{\left(\frac{-k_x^2 - k_y^2 - k_z^2}{\mu\varepsilon}\right)} \qquad (6.78)$$

Returning to the wave equation, eqn. (6.3), and using an argument similar to that used to obtain eqns. (6.23) and (6.24), the resonant frequency can be

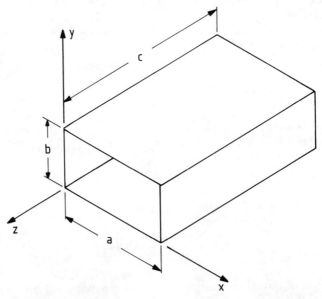

Figure 6.16 A rectangular resonant cavity showing its dimensions and its relationship to the rectangular coordinate system.

related to the dimensions of the cavity shown in Figure 6.16:

$$f_0 = \frac{1}{2}\sqrt{\left(\frac{(l/a)^2 + (m/b)^2 + (n/c)^2}{\mu\varepsilon}\right)} \qquad (6.79)$$

The mode of oscillation in the cavity is specified as TE_{lmn} or TM_{lmn}. The field pattern in the cavity varies in an integral number of half-wavelengths in each dimension.

A resonant cavity may also be constructed from a length of circular waveguide closed at each end with a short circuit. However, the circular waveguide cavity is not completely symmetrical in the same way as a rectangular waveguide cavity, so that it is not possible to produce a simple formula similar to eqn. (6.79) for the resonant frequency of a cylindrical cavity. The resonant frequency must be calculated from the cut-off conditions of the relevant circular waveguide modes. In order to simplify the calculation of the resonant frequency of a cylindrical cavity, a mode chart has been constructed relating the resonant frequency of a cylindrical cavity to its dimensions. The cavity is shown diagrammatically in Figure 6.17 and the mode chart in Figure 6.18. A third number is added to the mode nomenclature to denote the length in half-wavelengths of the equivalent circular waveguide. Otherwise, the mode nomenclature is the same as that used for the modes in circular waveguide. Then the cylindrical cavity modes are TM_{nml} or TE_{nml}. It is interesting to note that there are TM-modes with no variation of the fields along the length of the cavity.

The cavity may be coupled to an associated transmission line by a coaxial line terminating in an electric probe or a magnetic probe as shown in Figure 6.19. Alternatively it may be coupled to an adjacent waveguide by a small

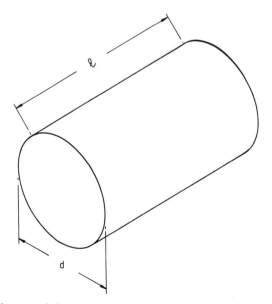

Figure 6.17 The shape and dimensions of a right circular cylindrical cavity.

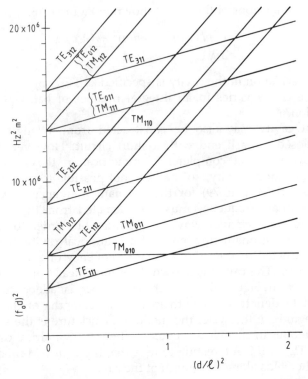

Figure 6.18 The mode chart for a right circular cylindrical resonator, diameter d and length l (see p. ix). (Reproduced with permission from Baden Fuller, *Microwaves* (3rd edn), © 1990 Pergamon Press Ltd.)

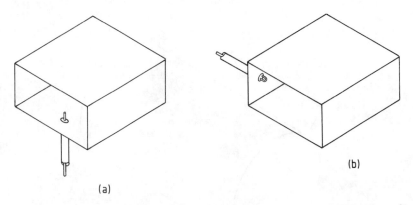

Figure 6.19 (a) An electric probe and (b) a magnetic probe coupling to the relevant field of the TE_{10n}-mode in a rectangular cavity (see p. ix).

aperture in one of the walls. It is found at most frequencies that there is very little penetration of the electromagnetic fields into the cavity and the cavity appears to be a short circuit at the end of the transmission line. If it is aperture coupled to the side of a waveguide, the cavity has very little effect on the fields in the waveguide. At the resonance frequencies, the cavity accepts the power from the transmission line or it abstracts appreciable power from the waveguide. The electromagnetic fields inside the cavity are of the same order as the fields in the rest of the system.

For many purposes, the design of resonant cavities is concerned with achieving high Q-factors and precise control of the resonant frequency. As usually a dielectric material dominates the attenuation in any transmission line, the highest Q-factors are provided by hollow metal cavities. Then the losses are controlled by the resistivity of the material of the cavity walls. Since the losses are proportional to the area of the walls but the stored power is proportional to the volume of the cavity, electrically large cavities have large Q-factors. However, large cavities can support a number of waveguide modes at different frequencies and it is necessary to determine all the possible resonant frequencies of any cavity. Rectangular cavities are easy to analyse but difficult to make to precise dimensions. Circular cavities are easy to make to precise dimensions and have higher Q-factors than rectangular cavities of the same size. It is usually necessary to design a cavity so that it supports only one mode over the proposed operating frequency band.

A cylindrical disc of low-loss high-permittivity material also acts as a resonant cavity. The modes are similar to those in circular waveguide but not the same because the cavity does not have any metallic walls. They are of high Q-factor and because of the high permittivity they are physically small. They are useful in conjunction with microstrip microwave circuits.

SUMMARY

The Q-factor of a resonant cavity is obtained from the width of its resonance curve at the half-power point:

$$Q = \frac{f_0}{\delta f} = 2\pi f_0 \frac{\text{energy stored}}{\text{power dissipated}} \qquad (6.74)$$

A simple resonant cavity consists of a length of transmission line with a short circuit at each end. The resonant frequency is

$$f_0 = \frac{nv}{2z} \qquad (6.77)$$

A rectangular resonant cavity has the resonant frequency

$$f_0 = \frac{1}{2}\sqrt{\left(\frac{(l/a)^2 + (m/b)^2 + (n/c)^2}{\mu\varepsilon}\right)} \qquad (6.79)$$

The resonant frequency of a cylindrical cavity is determined from the mode chart shown in Figure 6.16.

192 HOLLOW METAL WAVEGUIDE

High-Q cavities generally are large so they are able to support a number of different modes of oscillation in the cavity. Each mode gives rise to a different resonant frequency.

Example 6.6 Sketch the fields for the TM_{011}-mode in a cylindrical cavity and indicate the best position for (a) an electric probe, (b) a magnetic probe and (c) a waveguide aperture coupling to the magnetic field of an adjacent waveguide, to couple to the fields in the cavity.

Answer The fields are obtained by using a half-wavelength of the waveguide mode described by eqn. (6.74). The field pattern in the transverse cross-section of the cavity at the centre of the cavity is shown in Figure 6.13. The longitudinal electric field is a maximum along the axis of the cavity at each end. Therefore: (a) the best position for the electric probe, is in the centre of the circular end of the cavity; (b), (c) the magnetic probe and the aperture should be where the magnetic field parallel to the wall is a maximum, which is in the curved section of the cavity wall at the centre of the length of the cavity.

PROBLEMS

6.6 An air-spaced transmission line one metre long has a short circuit at each end. Determine its resonant frequencies.

[$150n$ MHz]

6.7 A rectangular cavity has the dimensions 1 cm by 2 cm by 3 cm. Find the first five resonant frequencies and identify the mode for each one.

TE_{011}	TE_{012}	TE_{013}	TE_{021}	TE_{014}
0.90	1.04	1.14	1.17	1.25 GHz

6.8 A cylindrical cavity of 3 cm diameter is either 2 cm long or 6 cm long. Determine the first six resonant frequencies and identify the mode for each one for each of the two cavities.

TM_{010}	TE_{111}	TM_{011}	TE_{211}	TM_{110}	TE_{112}
0.76	0.80	0.94	1.12	1.19	1.27 GHz

TE_{111}	TM_{010}	TE_{112}	TM_{011}	TM_{012}	TE_{211}
0.62	0.76	0.76	0.80	0.91	1.00 GHz

─── CHAPTER 7 ───────────────────────────

Optical Fibre

Aims: to give a theoretical and practical description of the optical fibre used as a dielectric rod waveguide for low-loss propagation at optical frequencies.

7.1 SURFACE WAVE

At optical frequencies, the dimensions of hollow metal waveguide become impossibly small. Light has a characteristic wavelength of about 1 μm so that any waveguide needs to have a cross-section of about 1 μm. Glass fibre can be reduced to these dimensions and is now used as a transmission medium for optical signals. Before analysing the modes that can propagate in a circular fibre, the mathematically simpler modes that can propagate parallel to a plane dielectric surface and in a plane thin film are discussed.

Consider a plane surface between two dielectric media similar to that shown in Figure 4.3. In this situation, the incident wave is at a grazing angle to the surface giving a direction of polarisation parallel to the surface. There are fields in the two media propagating at the same speed parallel to the surface and related by the boundary conditions. The fields for a TE-mode surface wave are shown in Figure 7.1. z is the direction of propagation. It is a form of plane wave in that there is no variation of any of the fields in the y-direction. There is a variation of the fields in the x-direction perpendicular to the surface and the fields decay to zero an infinite distance away from the surface. First consider the TE-mode whose fields are shown in Figure 7.1.

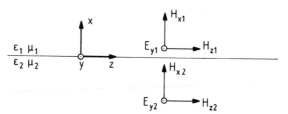

Figure 7.1 The fields of a TE-mode surface wave on the plane boundary between two dielectric media for propagation in the z-direction.

OPTICAL FIBRE

The wave equation is similar to eqn. (6.4) except that there is no variation of the fields in the y-direction. Therefore

$$\frac{\partial^2 H_z}{\partial x^2} + \frac{\partial^2 H_z}{\partial z^2} = -\omega^2 \mu \varepsilon H_z \tag{7.1}$$

Proceeding to solve this equation by the separation of variables gives a solution similar to eqn. (6.8),

$$k_x^2 + k_z^2 = -k^2 \tag{7.2}$$

If β is the phase constant for propagation in the z-direction then, similar to eqn. (6.12),

$$\beta = jk_z = \pm\sqrt{(k^2 + k_x^2)} \tag{7.3}$$

The z-dependent relationship is found from the solution to an equation similar to eqn. (6.7),

$$\frac{\partial^2 H_z}{\partial z^2} = k_z^2 H_z$$

to be similar to eqn. (4.21),

$$H_z = f(x) \exp[j(\omega t - \beta z)] \tag{7.4}$$

The field strength variation in the x-direction is found from the other equation similar to eqn. (6.7),

$$\frac{\partial^2 H_z}{\partial x^2} = k_x^2 H_z$$

The solution is of the form

$$H_z = f(t, z) \exp(\pm k_x x) \tag{7.5}$$

Equation (7.5) describes an exponential growth or decay of the field in the x-direction. Since the fields have to be zero at an infinite distance from the surface, eqn. (7.5) must describe an exponential decay away from the surface. Since it is an exponential decay, let $k_x = \alpha$.

However, it is found that there is a standing wave in the x-direction in medium 2 because the equivalent value of k_x^2 is negative. Then, from eqn. (7.3),

$$k_{x1} = \alpha_1 = \sqrt{(\beta^2 - k_1^2)} = \sqrt{(\beta^2 - \omega^2 \mu_1 \varepsilon_1)}$$
$$jk_{x2} = \alpha_2 = \sqrt{(k_2^2 - \beta^2)} = \sqrt{(\omega^2 \mu_2 \varepsilon_2 - \beta^2)} \tag{7.6}$$

Because it is describing the different parts of the same wave, β is the same in each medium. Therefore

$$\beta^2 = \omega^2 \mu_1 \varepsilon_1 + \alpha_1^2 = \omega^2 \mu_2 \varepsilon_2 - \alpha_2^2$$

Then the fields in the two dielectric media are given to

$$H_{z1} = H_1 \exp(-\alpha_1 x) \exp[j(\omega t - \beta z)] \qquad (7.7)$$

$$H_{z2} = H_2 \cos(\alpha_2 x + \phi) \exp[j(\omega t - \beta z)] \qquad (7.8)$$

where the phase ϕ of the standing wave in medium 2 is determined by the boundary conditions.

In order to apply the boundary conditions at the surface, it is necessary to obtain expressions for the components of the fields in the two media. Maxwell's curl equations in their component parts and including the z and t variation given in eqn. (7.4) are given in eqns. (6.28)–(6.33). Incorporating the new condition that

$$\frac{\partial}{\partial y} = 0$$

gives

$$j\beta E_y = -j\omega \mu H_x \qquad (7.9)$$

$$-j\beta E_x - \frac{\partial E_z}{\partial x} = -j\omega \mu H_y \qquad (7.10)$$

$$\frac{\partial E_y}{\partial x} = -j\omega \mu H_z \qquad (7.11)$$

$$j\beta H_y = j\omega \varepsilon E_x \qquad (7.12)$$

$$-j\beta H_x - \frac{\partial H_z}{\partial x} = j\omega \varepsilon E_y \qquad (7.13)$$

$$\frac{\partial H_y}{\partial x} = j\omega \varepsilon E_z \qquad (7.14)$$

Equation (7.9) gives

$$H_x = -\frac{\beta}{\omega \mu} E_y$$

which, when substituted into eqn. (7.13), gives

$$E_y = -\frac{j\omega \mu}{k_x^2} \frac{\partial H_z}{\partial x} \qquad (7.15)$$

$$H_x = -\frac{j\beta}{k_x^2} \frac{\partial H_z}{\partial x} \qquad (7.16)$$

Similarly, eqn. (7.12) gives

$$H_y = \frac{\omega\varepsilon}{\beta} E_x$$

which, when substituted into eqn. (7.10), gives

$$E_x = \frac{j\beta}{k_x^2} \frac{\partial E_z}{\partial x} \qquad (7.17)$$

$$H_y = \frac{j\omega\varepsilon}{k_x^2} \frac{\partial E_z}{\partial x} \qquad (7.18)$$

From eqns. (7.15)–(7.18) it is seen that the possible waves consist of a TE-mode with the fields E_y, H_x and H_z and a TM-mode with the fields E_x, E_z and H_y. From eqns. (7.15) and (7.16) the components of the fields of the TE-mode are given by eqns. (7.7) and (7.8) and substituting for α_1 and α_2 from eqn. (7.6),

$$E_{y1} = \frac{j\omega\mu_1}{\alpha_1} H_1 \exp(-\alpha_1 x) \exp[j(\omega t - \beta z)] \qquad (7.19)$$

$$H_{x1} = \frac{j\beta}{\alpha_1} H_1 \exp(-\alpha_1 x) \exp[j(\omega t - \beta z)] \qquad (7.20)$$

$$E_{y2} = -\frac{j\omega\mu_2}{\alpha_2} H_2 \sin(\alpha_2 x + \phi) \exp[j(\omega t - \beta z)] \qquad (7.21)$$

$$H_{x2} = -\frac{j\beta}{\alpha_2} H_2 \sin(\alpha_2 x + \phi) \exp[j(\omega t - \beta z)] \qquad (7.22)$$

At the boundary between the two media, the tangential components of the fields are equal and $x = 0$. Therefore

$$E_{y1} = E_{y2} \qquad (7.23)$$

$$H_{z1} = H_{z2} \qquad (7.24)$$

Substituting from eqns. (7.7) and (7.8) into eqn. (7.24) gives

$$H_1 = H_2 \cos\phi \qquad (7.25)$$

Then substituting from eqns. (7.19) and (7.21) into eqn. (7.23) gives

$$\frac{\mu_1}{\alpha_1} H_1 = -\frac{\mu_2}{\alpha_2} H_2 \sin\phi \qquad (7.26)$$

Therefore from eqns. (7.25) and (7.26),

$$\tan\phi = -\frac{\mu_1 \alpha_2}{\mu_2 \alpha_1} \qquad (7.27)$$

which gives the phase lag in the transverse plane due to the boundary for the TE-mode.

The properties of the TM-mode are obtained similarly. The wave equation is similar to eqn. (7.1) for E_z. Its solution is similar to eqns. (7.6) and (7.7),

$$E_{z1} = E_1 \exp(-\alpha_1 x) \exp[j(\omega t - \beta z)] \tag{7.28}$$

$$E_{z2} = E_2 \cos(\alpha_2 x + \phi) \exp[j(\omega t - \beta z)] \tag{7.29}$$

The other components of the fields are obtained by substitution into eqns. (7.17) and (7.18). Therefore

$$E_{x1} = -\frac{j\beta}{\alpha_1} E_1 \exp(-\alpha_1 x) \exp[j(\omega t - \beta z)] \tag{7.30}$$

$$H_{y1} = -\frac{j\omega\varepsilon_1}{\alpha_1} E_1 \exp(-\alpha_1 x) \exp[j(\omega t - \beta z)] \tag{7.31}$$

$$E_{x2} = \frac{j\beta}{\alpha_2} E_2 \sin(\alpha_2 x + \phi) \exp[j(\omega t - \beta z)] \tag{7.32}$$

$$H_{y2} = \frac{j\omega\varepsilon_2}{\alpha_2} E_2 \sin(\alpha_2 x + \phi) \exp[j(\omega t - \beta z)] \tag{7.33}$$

At the boundary between the two media, $x = 0$ and the tangential components of the fields are equal. Therefore

$$E_{z1} = E_{z2} \tag{7.34}$$

$$H_{y1} = H_{y2} \tag{7.35}$$

Substituting from eqns. (7.28) and (7.29) into eqn. (7.34) gives

$$E_1 = E_2 \cos\phi \tag{7.36}$$

and, substituting from eqns. (7.31) and (7.33) into eqn. (7.35) gives

$$\frac{\varepsilon_1}{\alpha_1} = -\frac{\varepsilon_2}{\alpha_2} \sin\phi \tag{7.37}$$

Therefore from eqns. (7.36) and (7.37),

$$\tan\phi = -\frac{\varepsilon_1 \alpha_2}{\varepsilon_2 \alpha_1} \tag{7.38}$$

which gives the phase change in the transverse plane due to the boundary for the TM-mode.

Mathematically, a wave has been described which propagates along the boundary between the two media. In the dimension perpendicular to the boundary, there is a standing wave of field in the one medium and an exponential decay of field in the other as shown in Figure 7.2. Effectively there is a standing wave in medium 2 but the direction of propagation is parallel to the surface between the two media. From a consideration of the components of the fields in the two media and by a solution of Maxwell's equations we have derived the condition of total internal reflection at a surface described in Section 4.4 by eqns. (4.60) and (4.61).

The condition of a standing wave in a medium of infinite extent is not a satisfactory description for a real structure but the theory is easily extended to describe parallel-sided dielectric waveguide.

SUMMARY

A surface wave propagates along the boundary between two dielectric media. There is an exponential decay of the field strengths in the direction perpendicular to the surface in the medium of lower permittivity. There is a standing wave in the medium of higher permittivity. The boundary supports a TE-mode surface wave with the transverse phase constant given by

$$\tan \phi = - \frac{\mu_1 \alpha_2}{\mu_2 \alpha_1} \qquad (7.27)$$

and the components of the fields in the two media are given by eqns. (7.6), (7.7) and (7.19)–(7.22).

The boundary also supports a TM-mode surface wave with the transverse phase constant given by

$$\tan \phi = - \frac{\varepsilon_1 \alpha_2}{\varepsilon_2 \alpha_1} \qquad (7.38)$$

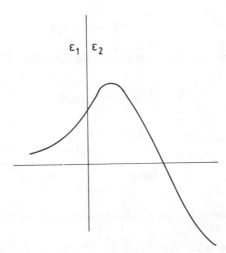

Figure 7.2 Showing the evanescent decaying field in the medium of lower permittivity.

7.2 DIELECTRIC FILM WAVEGUIDE

Consider a thin film of dielectric material on the surface of another dielectric material of smaller permittivity. The thickness of the film is of the same order as the wavelength of light. The film may be produced by processes similar to that used to manufacture semiconductor devices. A wave can propagate in the thin film provided that the angle of incidence of any ray is sufficiently small for total internal reflection to occur at both the faces of the film. Then waveguide modes propagate in the film similar to those modes that propagate in parallel plate waveguide as described in Section 5.1. Consider the thin film shown in Figure 7.3, where $\varepsilon_2 > \varepsilon_1 > \varepsilon_0$. The film usually consists of glass deposited onto a similar glass substrate of lower permittivity. For dielectric materials where the relative permeability is one, the index of refraction at the boundary n_{21} is defined by eqn. (4.45)

$$n_{21}^2 = \frac{\varepsilon_1}{\varepsilon_2} \tag{7.39}$$

From eqn. (4.59) the critical angle for total internal reflection is given by

$$\sin \psi_{2c} = n_{21}$$

However, the ray incidence is usually measured in terms of θ_2, therefore for total reflection at the air dielectric boundary the critical angle is given by

$$\cos \theta_{0c} = n_{20} \tag{7.40}$$

and for total reflection at the film substrate boundary it is given by

$$\cos \theta_{1c} = n_{21} \tag{7.41}$$

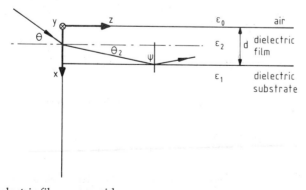

Figure 7.3 Dielectric film waveguide.

When $\theta_2 > \theta_{0c}$ there is partial reflection and refraction at both surfaces and the light ray is not trapped in a waveguide mode in the dielectric film. These are leaky modes which radiate out into space and are called *radiation modes*. Sometimes they are called *space radiation modes* or just *space modes*.

When $\theta_{1c} < \theta_2 < \theta_{0c}$, the wave is totally reflected at the air dielectric boundary but there is still partial reflection and refraction at the boundary between the film and the substrate. These are modes which leak into the substrate so that they are called *substrate radiation modes* or just *substrate modes*.

When $\theta_2 < \theta_{1c}$, the wave is totally reflected at both boundaries. There is wave propagation by waveguide modes in the thin film. These are called *guided film modes* or just *film modes*.

For the film modes, waveguide propagation occurs provided the incident wave generates a ray in the film such that $\theta_2 < \theta_{1c}$. From eqn. (7.41) the limiting angle is given by

$$\sin \theta_{1c} = \sqrt{(1 - n_{21}^2)} = \sqrt{\left(\frac{\varepsilon_2 - \varepsilon_1}{\varepsilon_2}\right)}$$

For the incident wave onto the front face of the film waveguide,

$$\sin \theta_c = \sqrt{\left(\frac{\varepsilon_2}{\varepsilon_0}\right)} \sin \theta_{1c} = \sqrt{\left(\frac{\varepsilon_2 - \varepsilon_1}{\varepsilon_0}\right)} = \sqrt{(n_2^2 - n_1^2)} \qquad (7.42)$$

The incident critical angle θ_c defines the angle within which an incident light ray will contribute to a propagating mode within the thin-film waveguide. This acceptance angle is called the *numerical aperture* (NA). Therefore

$$\text{NA} = \sin \theta_c = \sqrt{(n_2^2 - n_1^2)} = \sqrt{(\varepsilon_{r2} - \varepsilon_{r1})} \qquad (7.43)$$

where n_2 and n_1 are the refractive indices of the two dielectric materials and ε_{r2} and ε_{r1} are the corresponding relative permittivities.

The propagating waveguide mode is a fully defined mode propagating along a direction in the plane of the film with fields outside the propagating film having an exponential decay in the direction perpendicular to the plane of the film. Expressions for the fields in the wave are obtained from eqns. (7.19)–(7.22) for the TE-modes:

$$E_{y1} = \frac{j\omega}{\alpha_1} H_1 \exp[-\alpha_1 (x - d)] \exp[j(\omega t - \beta z)] \qquad (7.44)$$

$$E_{y2} = \frac{j\omega}{\alpha_2} H_2 \sin(\alpha_2 x + \phi_{20}) \exp[j(\omega t - \beta z)] \qquad (7.45)$$

$$E_{y0} = \frac{j\omega}{\alpha_0} H_0 \exp(\alpha_0 x) \exp[j(\omega t - \beta z)] \qquad (7.46)$$

and expressions for the other components of the fields are similar. The amplitude of the field components in the cross-section of the waveguide is shown in Figure 7.4 for one particular mode. The phase lag at the substrate boundary does not appear in eqn. (7.45) because it occurs in the evaluation of the constant H_2. As shown by Figure 7.4, any mode that can propagate in the thin-film waveguide has an integral number of half-wavelengths in the cross-sectional dimension less the phase lag at each boundary. Therefore

$$\alpha_2 d - \phi_{20} - \phi_{21} = m\pi \quad (7.47)$$

where m is any integer and ϕ_{20} and ϕ_{21} are given by expressions similar to eqn. (7.27). Therefore, substituting for ϕ_{20} and ϕ_{21},

$$\tan^{-1}\left(-\frac{\alpha_2}{\alpha_0}\right) + \tan^{-1}\left(-\frac{\alpha_2}{\alpha_1}\right) = \alpha_2 d - m\pi \quad (7.48)$$

where

$$\alpha_2 = \sqrt{(\omega^2 \mu_0 \varepsilon_2 - \beta^2)} \quad (7.49)$$
$$\alpha_1 = \sqrt{(\beta^2 - \omega^2 \mu_0 \varepsilon_1)} \quad (7.50)$$
$$\alpha_0 = \sqrt{(\beta^2 - \omega^2 \mu_0 \varepsilon_0)} \quad (7.51)$$

Take the tangent of both sides of eqn. (7.48),

$$-\frac{(\alpha_2/\alpha_0) + (\alpha_2/\alpha_1)}{1 - \alpha_2^2/(\alpha_0 \alpha_1)} = \tan \alpha_2 d \quad (7.52)$$

because $\tan m\pi = 0$.

For the TM-modes, the phase angles at the boundaries are given by eqn. (7.38). Therefore

$$-\frac{\varepsilon_0 \alpha_2/(\varepsilon_2 \alpha_0) + \varepsilon_1 \alpha_2/(\varepsilon_2 \alpha_1)}{1 - \varepsilon_0 \varepsilon_1 \alpha_2^2/(\varepsilon_2^2 \alpha_0 \alpha_1)} = \tan \alpha_2 d \quad (7.53)$$

For weakly guiding films, which are made using a glass film of a slightly different composition to the glass of the substrate, $\varepsilon_2 \approx \varepsilon_1$. There is only

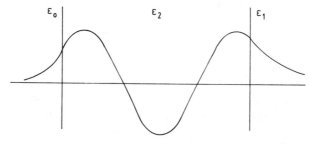

Figure 7.4 The amplitude of the field components in the cross-section of a dielectric film optical waveguide for one particular mode of propagation.

about 1% difference in the permittivities. The ray angle θ_2 is very small, $\alpha_0 \gg \alpha_1$ and $\alpha_0 \gg \alpha_2$. Therefore, for both types of modes,

$$-\frac{\alpha_2}{\alpha_1} = \tan \alpha_2 d \tag{7.54}$$

Define two new variables,

$$\left.\begin{array}{l} u = \alpha_2 d \\ v = \alpha_1 d \end{array}\right\} \tag{7.55}$$

Then substituting into eqn. (7.54) gives

$$-\frac{u}{v} = \tan u$$

or

$$v = -u \cot u \tag{7.56}$$

This equation gives one condition for any particular mode to propagate in the thin film. The other condition is derived from the definition of u and v:

$$u^2 + v^2 = (\alpha_2^2 + \alpha_1^2)d^2 = \omega^2 \mu_0 (\varepsilon_2 - \varepsilon_1) d^2$$

$$= \left(\frac{2\pi d}{\lambda}\right)^2 (n_2^2 - n_1^2) = V^2 \tag{7.57}$$

V is called the *generalised frequency parameter* of the film or just the *film parameter*. The solution of the two eqns. (7.56) and (7.57) gives the condition for a propagating mode to exist in the thin-film optical waveguide. The solution may be obtained numerically using a computer or, in the absence of a computer, the two expressions may be plotted in the u–v plane.

Equation (7.57) is the equation of a circle on the u–v plane of radius V. Substituting from eqn. (7.43) gives the film parameter in terms of the numerical aperture,

$$V = \frac{2\pi d}{\lambda} \sqrt{(n_2^2 - n_1^2)} = \frac{2\pi d}{\lambda} (NA) \tag{7.58}$$

u is called the transverse phase parameter and v is the transverse attenuation parameter.

According to eqn. (7.57), u cannot be greater than V and this defines the number of modes which can propagate in the thin-film waveguide. Because the cotangent function in eqn. (7.56) is periodic, varying between $\pm\infty$ in each π, there will be one solution to eqns. (7.56) and (7.57) in each $\pi/2$ for u. The relationship in eqn. (7.56) is plotted in Figure 7.5 and it is seen that

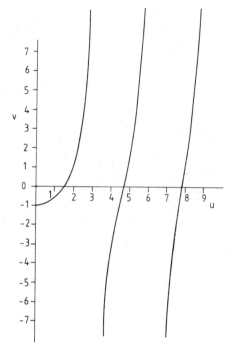

Figure 7.5 A plot of the relationship, $v = -u \cot u$.

there is only one solution to the equations for $u < \pi/2$. Therefore the total possible number of solutions to the equations is given approximately by

$$m = \frac{2V}{\pi} - 1 \tag{7.59}$$

rounded up to the nearest integer. However, each solution to the mathematical relationships represents both a TE_m and a TM_m mode, so that the total number of modes is $2m$. At the cut-off condition, the number of the mode is given by eqn. (7.59) exactly so that the cut-off of the mth mode is given by

$$V_m \approx (m+1)\frac{\pi}{2} \tag{7.60}$$

However, observation of Figure 7.6 shows that eqn. (7.60) is true only for odd-numbered modes. At cut-off, $v = 0$, therefore from eqn. (7.55) $\alpha_1 = 0$, and from eqn. (7.50)

$$\beta^2 = \omega^2 \mu_0 \varepsilon_1$$

Substituting back into eqns. (7.49) or (7.57) shows that

$$\alpha_2 = \frac{V}{d}$$

Therefore at cut-off of the guided film wave, there is a plane wave propagating in the substrate, an exponential decay of the fields in the film and no guidance provided by the film.

Thin-film waveguide may be further constrained into a propagating path in the plane of the film to become a rectangular dielectric waveguide supported on a dielectric substrate. The mathematical expressions for the fields in such a waveguide are much more complicated than those given in this section and will not be developed here. The process of defining the shape of the waveguide is no more difficult than that commonly used in the manufacture of integrated circuits, and optical circuits can be produced on dielectric substrates. However, the main object of the mathematical development given in this section is to present the concept of modes in dielectric waveguide. The thin-film waveguide described here is little used but it provides a simple mathematical introduction to the mathematically much more complicated optical fibre described in the next section.

SUMMARY

Waveguide modes of propagation are supported in a dielectric film on the surface of a dielectric substrate of smaller permittivity. The incident critical angle is the angle within which an incident light ray contributes to a propagating mode within the thin-film waveguide:

$$\text{numerical aperture} = NA = \sin\theta_c$$
$$= \sqrt{(n_2^2 - n_1^2)} = \sqrt{(\varepsilon_{r2} - \varepsilon_{r1})} \tag{7.43}$$

The variation in field strength for the different components of the wave are given by eqns. (7.44)–(7.46).

The propagating condition for any particular mode is given by

$$v = -u \cot u \tag{7.56}$$

where

$$u = \alpha_2 d \quad \text{and} \quad v = \alpha_1 d \tag{7.55}$$

The generalised frequency parameter or film parameter is V, given by

$$V^2 = u^2 + v^2 \tag{7.57}$$

where

$$V = \frac{2\pi d}{\lambda} \sqrt{(n_2^2 - n_1^2)} = \frac{2\pi d}{\lambda} (NA) \tag{7.58}$$

and d is the thickness of the film. The relationship in eqn. (7.56) is plotted in Figure 7.5 so that values of u and v appropriate to each mode of propagation may be found by drawing on Figure 7.5 the circle radius V appropriate to eqn. (7.57). There is a TE and a TM-mode associated with each solution to eqns. (7.56) and (7.57). The number of possible pairs of modes is given approximately by

$$m = \frac{2V}{\pi} - 1 \qquad (7.59)$$

Example 7.1 A thin-film waveguide consists of a film of glass of refractive index 1.48 on the surface of a glass substrate of refractive index 1.46. The film is of thickness 5 μm. Light at a frequency of 2.3×10^{14} Hz is propagating in the film waveguide. Calculate the free-space wavelength of the light, the film parameter, the number of modes that will propagate, and the phase propagation constants for each mode. Sketch the electric field strength variation in the cross-section of the waveguide for each propagating mode.

Answer The free-space wavelength is the characteristic wavelength given by eqn. (4.30):

$$\lambda = \frac{c}{f} = \frac{3.0 \times 10^8}{2.3 \times 10^{14}} = 1.300 \ \mu m$$

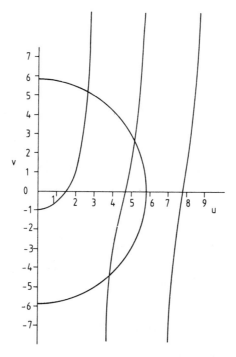

Figure 7.6 Part of the answer to Example 7.1. A circle of radius $V = 5.86$ plotted on the graph of Figure 7.5.

The film parameter is given by eqn. (7.58). Therefore

$$V = 5.86$$

The number of modes is given approximately by eqn. (7.59). Therefore

$$m = 2.73$$

and three pairs of modes are able to propagate in the film waveguide. Further information on the modes is obtained by plotting a circle of radius $V = 5.86$ on Figure 7.5. This solution is given in Figure 7.6 and the results read from the graph are given in Table 7.1. β is calculated from eqns. (7.49) and (7.55).

Table 7.1

u	v	β
5.16	2.80	7.08×10^6
3.90	−4.40	7.11×10^6
2.78	5.25	7.13×10^6

For comparison, the phase constant for a plane wave propagating in an infinite medium of the glass of the film is $\beta = 7.15 \times 10^6$ m^{-1}. Plots of the field strength variation of the three modes are given in Figure 7.7.

PROBLEMS

7.1 A thin-film waveguide has indices of refraction of 1.51 and 1.50. The external medium is air. Calculate the numerical aperture of this waveguide and the total angle of incidence onto the front face of the waveguide so that total internal reflection occurs inside the waveguide.

[0.1735, 10°]

7.2 The thin-film waveguide of Problem 7.1 is 7 μm thick. Calculate the value of the film parameter and find the number of modes that can propagate in the waveguide for light of frequency 2.0×10^{14} Hz.

[5.09, 3]

7.3 CIRCULAR FIBRE

The simplest form of optical waveguide and the easiest to construct is the optical fibre. It consists of a very small circular dielectric rod waveguide made out of glass or plastic. For the optical waveguide to operate in the dominant mode, its diameter needs to be of the same order as the characteristic wavelength of the light, that is, a few micrometres. In order to have a

CIRCULAR FIBRE 207

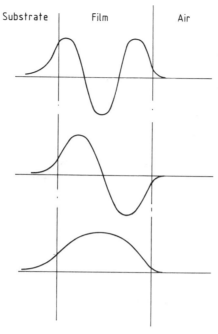

Figure 7.7 Another part of the answer to Example 7.1. Comparative field strength graphs for the three modes propagating in the thin-film waveguide.

fibre of a more reasonable size, optical fibres are made in the form of a cladded core fibre where the thin dielectric core waveguide is surrounded by an outer sheath of dielectric medium having a lower permittivity. In terms of wavelength, the outer diameter of the sheath is so large that its outer boundary may be ignored. For the purposes of analysis, the optical fibre may be considered as a circular dielectric rod of permittivity ε_1, refractive index n_1, in a surrounding medium of permittivity ε_2, refractive index n_2, where $\varepsilon_1 > \varepsilon_2$. The incident ray is very close to the axis of the fibre so that total internal reflection occurs at the boundary between the two media, as shown in Figure 7.8. As long as $\theta < \theta_c$, where θ_c is given by eqns. (7.42) and (7.43), the ray will be totally internally reflected and the *numerical aperture* for the circular fibre is the same as that for the dielectric film waveguide.

The circular dielectric rod waveguide is best analysed in terms of cylindrical polar coordinates, as shown in Figure 6.7. The circular boundary at radius $r = a$ on that diagram is the boundary between two dielectric materials. Expressed in cylindrical polar coordinates, the wave equation is given by eqn. (6.45) and its solution by eqn. (6.51), which is repeated here for completeness:

$$E_z = [AJ_n(k_c r) + BY_n(k_c r)] \exp[j(\omega t - n\theta - \beta z)] \qquad (7.61)$$

where A and B are arbitrary constants and J_n and Y_n are Bessel functions of order n of the first and second kind respectively. Values of the two

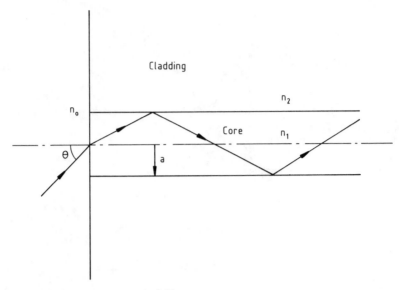

Figure 7.8 A light ray in an optical fibre.

lowest-order Bessel functions are given in Figure 6.8. The wave equation in terms of the longitudinal component of the magnetic field is given by eqn. (6.46) and its solution is similar to eqn. (7.61):

$$H_z = [CJ_n(k_c r) + DY_n(k_c r)] \exp[j(\omega t - n\theta - \beta z)] \qquad (7.62)$$

Mathematically, the boundary conditions for the dielectric rod waveguide are much more complicated than those for the hollow metal waveguide. It is necessary to obtain expressions for all the components of both the fields in each medium and then to equate the values of the tangential components of the fields on each side of the boundary at the boundary at $r = a$. It is then found that the equations for the conditions at the boundary produce a relationship between E_z and H_z, and it is no longer possible to separate the modes of propagation simply into transverse electric and transverse magnetic (TE and TM) modes. Except for the $n = 0$ modes, where separation into TE and TM-modes is still possible, all other modes have longitudinal components of both the electric and magnetic field. They can no longer be classified simply into transverse electric and transverse magnetic modes. They are classified as HE- or EH-modes, the first letter denoting the larger longitudinal component of the field. Therefore, HE-modes are quasi TE-modes and EH-modes are quasi TM-modes. The mode numbers are the same as those used in the descriptions of eqns. (6.67) and (6.70). The first mode number is the number n occurring in eqns. (7.61) and (7.62), and denotes the number of oscillations of the field quantities as we move in a circle around the centre of the waveguide. The second mode number denotes the number of oscillations of the field quantities in the radial direction.

Using the expressions for E_z and H_z in eqns. (7.61) and (7.62), expressions for the other components of the fields may be obtained by substitution into eqns. (6.63)–(6.66). In the core k_c^2 is positive and is given by

$$k_{c1}^2 = \omega^2 \mu_0 \varepsilon_1 - \beta^2 \qquad (7.63)$$

In the cladding it is negative and is given by

$$k_{c2}^2 = \omega^2 \mu_0 \varepsilon_2 - \beta^2 \qquad (7.64)$$

because $\beta^2 > \omega^2 \mu_0 \varepsilon_2$. In the core, $Y_n = \infty$ at $r = 0$ so that the arbitrary constants B and D have to be zero for the fields to exist, and the longitudinal components of the fields in the core are given by eqns. (6.54) and (6.68). In the cladding k_c is complex giving rise to a modified version of Bessel's equation, eqn. (6.49). Its solution is in terms of *modified Bessel functions*:

$$E_z = [PI_n(k_{c2}r) + QK_n(k_{c2}r)] \exp[j(\omega t - n\theta - \beta z)] \qquad (7.65)$$

where P and Q are arbitrary constants and I_n and K_n are the modified Bessel functions of the first and second kind respectively. I_n is a steadily increasing function of r whereas K_n is a steadily diminishing function of r. As the fields in any practical dielectric rod waveguide mode ought to decay to zero outside the rod as the radial distance increases, the constant P has to be zero and the longitudinal component of the electric field in the cladding is given by

$$E_z = [QK_n(k_{c2}r)] \exp[j(\omega t - n\theta - \beta z)] \qquad (7.66)$$

For the modes of propagation in the dielectric rod waveguide, the field quantities inside the rod vary radially according to a Bessel function relationship similarly to those of a similar mode in circular waveguide. In the cladding they decay with radial distance according to the modified Bessel function relationship which is very similar to an exponential decay. There are no simple mathematical expressions for the cut-off conditions in dielectric rod waveguide similar to eqns. (6.53) and (6.69) for circular waveguide. However, the relationships have been solved and a mode chart may be plotted giving the relationship between the phase constant, β, and the dimensions of the waveguide in the form of the normalised frequency, except that the normalised frequency is now defined in terms of the *radius* of the dielectric rod:

$$V = \frac{2\pi a}{\lambda} \sqrt{(n_1^2 - n_2^2)} = a\sqrt{(k_1^2 - k_2^2)} \qquad (7.67)$$

The mode chart for the dielectric rod waveguide is shown in Figure 7.9. There is no cut-off for the dominant HE_{11}-mode and this mode is able to propagate down to the lowest frequencies. The cut-off for the next mode is

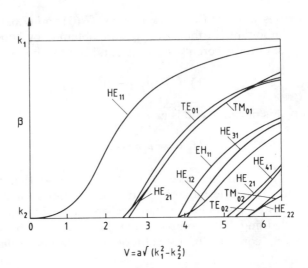

Figure 7.9 A mode chart for the dielectric rod waveguide of radius a (see p. ix). (Reproduced by permission from Halley. *Fibre Optic Systems*, © 1987 John Wiley & Sons Ltd.)

$V = 2.405$, so that the dominant mode is able to propagate alone provided that $V < 2.405$. For values of V greater than this the number of modes increases rapidly. The number of modes is given approximately by

$$m \approx \tfrac{1}{2} V^2 \qquad (7.68)$$

Below cut-off the mode propagates radially and is lost in the cladding. The electric and magnetic field patterns for the dominant mode in dielectric rod waveguide are similar to those for the dominant TE_{11}-mode in circular waveguide. The electric and magnetic field pattern in the cross-section of the dielectric rod is shown in Figure 7.10. One big difference compared with circular metallic waveguide is that the fields extend outside the dielectric rod boundary into the cladding. Single-mode fibres are possible at a reasonable radius provided $n_1 \approx n_2$, that is for a weakly guiding fibre.

Single-mode fibres are the most efficient, that is they have the lowest loss, but they are also the most difficult to manufacture because of their small size. They are used as a long-distance communication medium. For short distances, multi-mode larger-diameter fibres are usually more convenient.

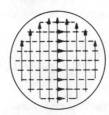

Figure 7.10 The field pattern of the dominant mode in the transverse cross-section of dielectric rod waveguide: ———→ electric field; - - - - → magnetic field.

For multi-mode propagation, the phase velocity and the group velocity for each mode is different and any signal becomes distorted.

The process of manufacture of the fibre involves diffusion of various impurities into the glass to modify its permittivity and further processing in the near molten state. At high temperature there is further diffusion of the various impurities between the glass of the core and the cladding and the step change in permittivity at the boundary becomes blurred. The ideal step change in permittivity or refractive index at the boundary is very difficult to make. Most fibres actually have a smooth transition between the core and the cladding and, provided that the change in refractive index occurs in a distance that is short compared with the characteristic wavelength, the step index core analysis is valid. If the distance of change is greater than the wavelength or if the whole core has a graded index, then a further more complicated analysis is necessary. These are the *graded index fibres*. The graded index fibre performs better than the step index fibre for multi-mode propagation because the mode providing an off-axis ray travels faster than one providing an on-axis ray and tends to keep up with it. The *parabolic index fibre* has a parabolic variation of refractive index across the whole cross-section of the fibre and has the following advantages. It minimises delay distortion for multi-mode propagation. The core is twice as large as the core of a step index fibre for the same number of modes and a fibre with a larger light-carrying cross-section simplifies the process of joining fibres together or of connecting them to light sources or detectors. A summary of the effect of the different fibres is shown in Figure 7.11, in terms of increasing efficiency of performance. A glass preform is made having the required cross-sectional variations in permittivity which is about 1 m long and 20 mm in diameter. It is then drawn into a thin fibre 125 μm in diameter and a few kilometres long. That is then sheathed in an opaque polymer protective jacket. The core depends on the type of fibre. It is as small as 5 μm for single-mode fibre and may be 50 μm diameter or larger for a multi-mode fibre. The outer diameter of the cladding is usually between 100 to 125 μm in diameter.

Cheaper less critical systems can make use of plastic fibres for communication over small distances. They are multi-mode graded index fibres. They have a larger diameter than glass because their permittivity is lower. This means that they are easier to handle and can be used with cheaper light sources and detectors. Jointing between fibres is also easier. According to theory, the attenuation in plastic fibres cannot be less than about 1.5 dB/km. In practice the loss is considerably greater than this and the high loss limits the use of plastic fibres to communication links of 200 m or less. However, 200 m is sufficient for many office systems.

SUMMARY

For a step index circular rod fibre waveguide, the acceptance angle is the numerical aperture given in eqn. (7.43).

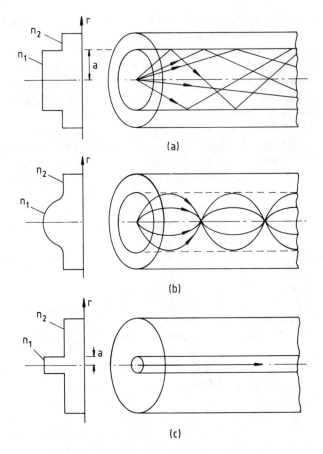

Figure 7.11 Optical fibre construction: (a) multi-mode step index fibre; (b) multi-mode parabolic index fibre; (c) single-mode step index fibre.

The field quantities in the core of the waveguide are governed by *Bessel functions*:

$$E_z = [AJ_n(k_{c1}r)] \exp[j(\omega t - n\theta - \beta z)] \tag{7.61}$$

In the cladding they are governed by *modified Bessel functions*:

$$E_z = [QK_n(k_{c2}r)] \exp[j(\omega t - n\theta - \beta z)] \tag{7.66}$$

For the circular rod waveguide, the generalised frequency parameter is now defined in terms of the *radius* of the core:

$$V = \frac{2\pi a}{\lambda} \sqrt{(n_1^2 - n_2^2)} \tag{7.67}$$

For a *step index fibre*, the propagating modes are given in Figure 7.9.

For multimode propagation, the *graded index* or particularly the *parabolic index* fibre gives the best performance.

CIRCULAR FIBRE

Example 7.2 A step index fibre is made of glass having the indices of refraction, $n_1 = 1.51$, $n_2 = 1.49$, and core diameter 6 μm. How many modes are able to propagate for light with a characteristic wavelength of 1.30 μm?

Answer The generalised frequency parameter of the fibre is given by

$$V = \frac{2\pi \times 3}{1.3} \sqrt{(1.51^2 - 1.49^2)} = 3.55$$

Using the approximate formula given in eqn. (7.68),

$$m = \tfrac{1}{2} V^2 = 6.3$$

so that the formula gives six modes. For a more accurate assessment reference is made to the graph in Figure 7.9, where it is seen that for $V = 3.55$, only four modes are able to propagate. They are the HE_{11}-, HE_{21}-, TE_{01}- and TM_{01}-modes.

Another two modes start propagating for $V > 3.832$, so the approximate formula is not very far wrong.

Example 7.3 A step index fibre is made of glass having a refractive index of 1.61. Find the relationship between the core diameter, $2a$, and the refractive index difference, Δn, for it to operate as a single-mode fibre with light of wavelength 1.55 μm.

Answer The condition for single-mode operation is $V < 2.405$. In terms of Δn, eqn. (7.67) may be written

$$V = \frac{2\pi a}{\lambda} \sqrt{[n_1^2 - (n_1 - \Delta n)^2]} = \frac{2\pi a}{\lambda} \sqrt{(2n_1 \Delta n)}$$

Therefore, for the limiting condition,

$$2a\sqrt{\Delta n} = 0.66 \ \mu m$$

This relationship between Δn and $2a$ is plotted in Figure 7.12.

PROBLEMS

7.3 A step index fibre has refractive indices of 1.51 and 1.50. The external medium is air. Calculate the numerical aperture for this waveguide. The core diameter is 7.0 μm. How many modes are able to propagate at a wavelength of 1.55 μm?

[0.1735, 7]

7.4 A step index fibre has refractive indices of 1.43 and 1.42. Calculate the limiting core diameter for single-mode operation at 1.30 μm wavelength.

[5.9 μm]

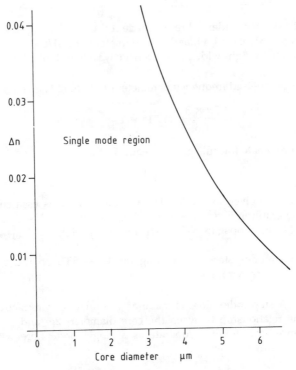

Figure 7.12 The answer to Example 7.3, showing the variation of Δn for single-mode operation versus core diameter for a step index fibre.

7.4 ATTENUATION AND DISPERSION

The physical length of the light transmission path that may be used without the need for signal amplification is governed by the absorption or attenuation of the electromagnetic wave. One of the advantages of the optical fibre as a telecommunication signal medium is its low attenuation, leading to the possibility of a transmission path of over 100 km without the need for any intermediate amplifiers. There are three physical mechanisms which contribute to the attenuation of the wave in an optical fibre.

Absorption loss. The electromagnetic wave is absorbed in the material of the optical waveguide. Materials are required which have a low absorption of light waves. Impurities contribute to the absorption so that special high-purity low-loss materials need to be used for the lowest attenuation fibres. The doping materials used to cause the change of permittivity for the core are impurities and can contribute to the attenuation.

Scattering loss. Light is deflected by inhomogeneities in the waveguide material and is scattered out through the cladding of the fibre. At the very short wavelengths this scattering is inevitable because it is caused by imperfections in the crystal structure of the glass of the fibre. This scattering is called *Rayleigh scattering*. Propagation of the electromagnetic wave through the material is largely undisturbed provided the wavelength of the wave is large compared to the size of the irregularities. The attenuation due to scattering increases as the wavelength approaches the dimensions of the irregularities in the crystal. Rayleigh scattering loss is also increased by the doping needed to control the permittivity of the core.

Sheath loss The cladding of the optical fibre is surrounded by an opaque protective sheath. As all the fields in the fibre theoretically extend to infinity, even though with a quasi-exponential decay, a small amount of electromagnetic power goes into the sheath. Fibres are usually designed so that sheath loss is negligible.

The total attenuation of a typical silica fibre is shown in Figure 7.13. The Rayleigh scattering provides a limit on the use of low-loss fibre at small wavelengths. This loss decreases as the fourth power of the wavelength. At the longer wavelengths there is an infrared absorption peak in silica materials. It is this infrared absorption which makes a glasshouse into a heat absorber. The glass windows transmit the light and ultraviolet radition into the glasshouse but absorb the lower temperature infrared radiation of the contents, so acting as an energy absorber. There is a low-loss region between the two attenuation peaks, which gives an absolute theoretical attenuation minimum for a silica fibre. It can be seen from the graph in Figure 7.13 why electromagnetic radiation of 1.30 and 1.55 μm wavelength is used for long-distance optical fibre communication systems. The attenuation peak at about 1.4 μm is due to water absorption, because it is difficult to eliminate water contamination completely during manufacture of fibres. Other contaminants give other absorption peaks for some fibres but these may all be eliminated by having the highest purity material. In the best fibres, the water absorption peak is the only significant departure from the

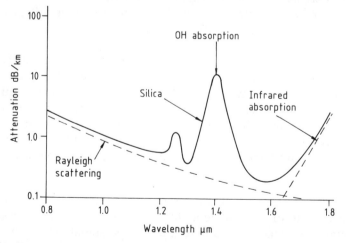

Figure 7.13 Attenuation in silica as used for optical fibres as a function of the free space optical wavelength.

theoretical minimum attenuation curve. The minimum attenuation with silica fibres is about 0.2 dB/km.

For even lower attenuation, materials having little or no infrared absorption or having the infrared absorption peak at a longer wavelength must be used in the region of the steadily reducing Rayleigh scattering with increase in wavelength. A fluoride glass has a theoretical attenuation limit of 0.03 dB/km at 2.5 μm wavelength. It is being investigated as a possible material for the next generation of low-loss optical fibres.

The largest contribution to attenuation in optical fibre systems are the joints between lengths of fibre. Demountable joints are called connectors. The ends of the fibre must be plane and square and located with high precision because any misalignment or gap leads to a large increase in attenuation. The ends of the fibre can be cleaved with such accuracy that polishing may not be necessary. Permanent joints are called splices. Accurately cleaved and located fibre ends may be fused together to make a continuous joint or may be fixed using a jointing compound having the same permittivity as the glass of the fibre.

The lowest loss single-mode fibres have the smallest diameter core, so those are the most difficult to splice or connect together. This is one of the advantages of using multi-mode graded index fibres having a much larger core diameter where the higher attenuation can be tolerated.

Dispersion is another cause of signal deterioration in optical fibre waveguides. Dispersion means that signals of different frequency travel at different speeds in the waveguide so that the received signal is a distorted version of the transmitted signal. There are three main causes of dispersion listed in order of increasing nuisance.

Material dispersion. The permittivity of any material varies with frequency. With most optical fibre glasses, the variation effect is small.

Waveguide dispersion. For any waveguide mode that has a cut-off frequency, the waveguide wavelength and the phase and group velocities vary with frequency as shown by eqns. (5.7), (5.9) and (5.14).

For both material and waveguide dispersion, any components of the signal that vary from the nominal carrier wave frequency arrive at the receiver at different times and signal distortion occurs. The received pulse width is broadened, so limiting the distance for intelligible communication. Cheaper light sources tend to generate relative broadband signals, so increasing the problems with dispersion. Both the material dispersion factor and the waveguide dispersion factor must be multiplied by the spectral width $\Delta \lambda$ of the light source, to give the total dispersion per unit length of optical fibre.

Modal dispersion. In any waveguide supporting a number of different modes, the group velocity of each mode is different and the signals carried in each mode arrive at the receiver at different times and signal distortion occurs. In normal waveguide systems, it is usual to operate so that only the dominant mode can propagate in the waveguide. This is monomode propagation. It is also the ideal in fibre optic systems but because of the necessary small size of the core it may not be the simplest or the cheapest for the shorter communication links. In multi-mode optical fibres many hundreds of modes may propagate and modal dispersion is the dominant process leading to signal deterioration. For the graded index, and particularly the parabolic index, multi-mode fibre, the modal dispersion is greatly reduced compared with the step index fibre having the same core diameter.

Modal dispersion is measured as the difference in propagation times of the modes with the slowest and fastest group velocities. However, for long fibres, the simple concept of modal dispersion is modified by mode conversion. In any multi-mode waveguide where the different modes are all propagating at approximately the same velocity, the fields due to the different modes interact and certain modes generate the fields of other modes. This is called mode conversion and reduces the damaging effect of modal dispersion. Up to a certain critical length, the coupling length, the effect of modal dispersion is linear, the effect increasing proportionally with length. However, for lengths greater than the critical length, the effect of modal dispersion is reduced and increases proportionally with the square root of length. In a long fibre, little of the energy propagates along the fibre in a single mode. Because most of the energy has propagated along different lengths of the fibre in different modes or in many different modes, the concept of a fastest or a slowest mode has little relevance. However, the dispersion of any propagated signal still increases as a function of the propagation distance.

Dispersion is defined as the difference between the longest and shortest propagation delays. The propagation delay per unit length is the reciprocal of the group velocity:

$$\frac{\tau}{l} = \frac{1}{v_g} = \frac{\partial \beta}{\partial \omega} \tag{7.69}$$

Then the dispersion is given by

$$\frac{\tau}{l} = \frac{\tau_{max} - \tau_{min}}{l} = \frac{1}{v_{g1}} - \frac{1}{v_{g2}} \tag{7.70}$$

The calculation of the modal dispersion for a long fibre is modified by mode conversion to be

$$\sigma_{modal} = \sigma_1[l_c + \sqrt{(l - l_c)}], \quad l \geq l_c \tag{7.71}$$

where l_c is the critical length for the onset of mode conversion and σ_1 is the modal dispersion per unit length measured on a short length of fibre for $l < l_c$.

Single-mode fibre can be designed so that the material dispersion is of opposite sign to the waveguide dispersion and they cancel out at a particular frequency. Such a fibre gives minimum dispersion. Because material dispersion and waveguide dispersion are both controlled by variations in the

218 OPTICAL FIBRE

frequency of the transmitted signal, the effects add linearly. However, modal dispersion in a multi-mode fibre is a completely independent phenomenon, and the two effects combine in an r.m.s. total. Therefore the total dispersion is given by

$$\sigma_{total}^2 = (\sigma_{material} + \sigma_{waveguide})^2 + \sigma_{modal}^2 \tag{7.72}$$

SUMMARY

There are three processes contributing to the total attenuation in an optical fibre;

absorption loss
scattering loss
sheath loss.

There are three processes contributing to the total dispersion distortion of the signal in an optical fibre:

modal dispersion
waveguide dispersion
material dispersion.

They combine as

$$\sigma_{total}^2 = (\sigma_{material} + \sigma_{waveguide})^2 + \sigma_{modal}^2 \tag{7.72}$$

Example 7.4 An optical fibre communication link is made of monomode fibre, 100 km long, having an attenuation of 0.3 dB/km. It is produced in cable lengths of 3 km and each splice introduces an additional loss of 0.2 dB. The connectors at each end introduce an additional loss of 1.0 dB per connector. What is the total loss, in dB, between the transmitter and the receiver? What is the power output into the receiver when the power input from the transmitter into the cable is 1.0 mW?

Answer The total cable consists of 34 individual lengths so there will be 33 splices. Therefore

total loss due to splices = 6.6 dB
total loss due to two connectors = 2.0 dB
total loss due to 100 km of cable = 30.0 dB

Therefore the total attenuation between transmitter and receiver is 38.6 dB. 38.6 dB is equivalent to a power ratio of 7244.
If the input is 1.0 mW, the output will be 0.138 μW.

Example 7.5 An optical fibre, 10 km long, has the following dispersion factors: the material dispersion is 6 ps/km nm, waveguide dispersion is 4 ps/km nm and the modal dispersion is 30 ps/km. The critical length for the onset of mode conversion is 800 m. The fibre propagates an electromagnetic wave from a laser having a spectral width $\Delta\lambda = 1.0$ nm. Calculate the total dispersion for 10 km of fibre.

Answer Both the material dispersion and the waveguide dispersion are multiplied by the length of the cable and the spectral width of the source so that

$$\sigma_{mat} = 60 \text{ ps}$$
$$\sigma_{wg} = 40 \text{ ps}$$

The total effect of the modal dispersion is given by eqn. (7.71), so that

$$\sigma_{modal} = 30(0.8 + \sqrt{9.2}) = 115 \text{ ps}$$

The total dispersion for the 10 km long fibre is given by eqn. (7.72):

$$\sigma = \sqrt{(100^2 + 115^2)} = 152.4 \text{ ps}$$

PROBLEMS

7.5 The cable described in Example 7.4 is used to construct a fibre-optic link over a distance of 50 km. Calculate the total attenuation through the cable and its connectors. It is connected to a transmitter giving 2.0 mW optical power. What is the optical power at the receiver?

[20.2 dB, 19 μW]

7.6 An optical fibre, 8.0 km long, has the following dispersion factors: the material dispersion is 5.0 ps/km nm, the waveguide dispersion is 4.0 ps/km nm and the modal dispersion is 2.0 ps/km. The critical length for the onset of mode conversion is 0.8 km. Calculate the total dispersion for 8 km of fibre, given that the optical transmitter is a light emitting diode with a spectral width $\Delta\lambda = 20$ nm.

[1.5 ns]

7.5 DIRECTIONAL COUPLER

In the optical waveguide, as in any other dielectric rod waveguide, the electromagnetic fields of the propagating wave extend outside the confines of the waveguide structure. The power in these peripheral fields is small but it is not negligible. In optical waveguides, coupling between waveguides is obtained by tapping these peripheral fields. If one waveguide structure is brought near to another waveguide so that their peripheral fields interact, electromagnetic energy is transferred from one waveguide to the other. A dielectric surface waveguide structure, similar to the dielectric film waveguide described in section 7.2, is shown in Figure 7.14. Here, the two waveguides are brought sufficiently close together for power transfer to occur over the length l. For optical couplers, the coupling is very weak but occurs over many wavelengths. For microwave waveguide, other methods of coupling may be used, giving quite tight coupling over a fraction of a wavelength, but the overall effect is the same. Power entering the coupling section at port 1 is gradually transferred to the other waveguide as it passes through the coupling section so that a proportion of the input comes out of port 3 and the remainder comes out of port 2. Provided that there are no reflections from the terminals of the device, nothing comes out of port 4.

In general waveguide terms, applicable to an electromagnetic wave of any frequency but not necessarily in the terminology of optical fibre systems, the performance of a directional coupler is usually quoted in terms of power ratio as

$$\text{coupling factor} = \frac{\text{power in port 3}}{\text{power in port 1}} \tag{7.73}$$

quoted in decibels. Because the practical directional coupler is not usually perfect, there is some unwanted power in the reverse arm which is measured as the directivity:

$$\text{directivity} = \frac{\text{power in port 4}}{\text{power in port 3}} \tag{7.74}$$

In the optical waveguide coupler, the power coupled per wavelength is very small but the coupling region is many wavelengths long, so that the

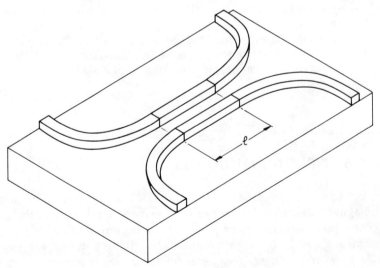

Figure 7.14 A dielectric surface waveguide directional coupler. l = interaction length.

total power coupled can be appreciable. In this kind of coupler, initially power is coupled into the empty waveguide. As the power increases in the secondary waveguide, some power is then coupled back into the primary waveguide. This means that, if the coupling region is sufficiently long, first, power is coupled into the secondary waveguide until at a certain distance all the power is transferred into the secondary waveguide, then, if the coupling continues, power is transferred back into the primary waveguide. If P_1, P_2 and P_3 are the powers in the relevant ports,

$$P_3 = P_1 \sin^2 \kappa l \qquad (7.75)$$

$$P_2 = P_1 \cos^2 \kappa l \qquad (7.76)$$

where κ is called the coupling coefficient and l is the interaction length as shown in Figure 7.14. Dielectric waveguide circuits can be produced on dielectric substrates using techniques similar to those used in the manufacture of integrated circuits.

Directional couplers can be made in optical fibres by removing some of the cladding in order to bring the cores into close proximity. The cross-section of the coupling region of such a coupler is shown in Figure 7.15. There are different methods of producing a cross-section similar to that shown in Figure 7.15. One is by bringing the fibres together, possibly twisting them together, and then heating them so that they fuse together. Another method, which possibly provides more control of the coupling factor, is to polish away some of the cladding of the fibre over the coupling length. The fibre is mounted in an arc, as shown in Figure 7.16, and a section of the cladding is removed by polishing. Then two such polished fibres may be brought together to give a cross-section similar to that shown in Figure 7.15 to make a directional coupler. If the two fibres are made to move sideways from one another, then the coupling coefficient varies and the device becomes a directional coupler with a variable coupling factor. The variable

Figure 7.15 Optical fibre directional coupler.

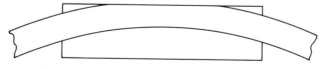

Figure 7.16 A polished fibre as used in a directional coupler.

coupler can also be used as a variable attenuator by having an absorber on port 3. A fixed attenuator can be made by replacing the second optical fibre with a slab of lossy dielectric. The amount of attenuation is controlled by the amount of power coupled out of the original optical fibre, therefore by the depth of cladding removed by polishing.

If a thin metal film is deposited on the polished surface of a monomode optical fibre, the fibre becomes polarisation sensitive. Referring to the field pattern of the dominant mode, shown in Figure 7.10, if the electric field is parallel to the metal film, the wave is completely reflected from the surface and no power is coupled out of the waveguide. Alternatively, if the electric field is perpendicular to the metal film, the wave is unaffected by the film and coupled normally into the space beyond the film. If the metal film lies along the flat shown between the two waveguide sections on Figure 7.15, the device becomes a polarisation-sensitive coupler. It separates the two possible perpendicular polarisations in the fibre. That polarisation having its electric field parallel to the metal film is transmitted through the device with minimum loss, whereas that polarisation having its electric field perpendicular to the metal film is coupled into the adjacent waveguide. Alternatively, if a wave with only one polarisation is required, the adjacent waveguide can be replaced by a slab of lossy dielectric on the other side of the metal film.

Just as a directional coupler may be used to split light signals into two or more separate paths, so they may also be used in reverse as combiners.

The general principles of directional couplers are applicable to any waveguide at any frequency. The couplers described in this section are those particularly used in optical communication systems.

SUMMARY

The directional coupler is used to couple a portion of the power from one waveguide into an adjacent waveguide. Its performance is given by

$$\text{coupling factor} = \frac{\text{power in port 3}}{\text{power in port 1}} \tag{7.73}$$

$$\text{directivity} = \frac{\text{power in port 4}}{\text{power in port 5}} \tag{7.74}$$

quoted in decibels.

The coupler operates by having another waveguide sited close to the main waveguide so that the peripheral fields of the two waveguides interact and electromagnetic power is transferred from one waveguide to the other. Optical fibres may be fused together or some of the cladding may be polished away. The power in the two output ports is given by

$$P_3 = P_1 \sin^2 \kappa l \tag{7.75}$$

$$P_2 = P_1 \cos^2 \kappa l \tag{7.76}$$

A polarisation-sensitive device is produced by depositing a thin metallic film onto the polished surface.

Example 7.6 What is the coupling factor of a half-power split directional coupler? Such a coupler has a coupling coefficient of 400 m^{-1}. What is its interaction length?

Answer The coupling factor is given by eqn. (7.73). The power ratio is 0.5, which is the same as 3 dB. The interaction length is given by eqn. (7.75). Therefore

$$\sin^2 \kappa l = \tfrac{1}{2}$$

$$\kappa l = \frac{\pi}{4}$$

Therefore

$$l = \frac{\pi}{4 \times 400} = 1.96 \text{ mm}$$

PROBLEMS

7.7 What is the power ratio of a 10 dB coupler? If its coupling coefficient is 300 m^{-1}, what is its interaction length?

[10:1, 1.07 mm]

7.8 A coupler has a coupling coefficient of 400 m^{-1}. What is the interaction length required to give 100% power transfer into the coupled waveguide? What happens if the interaction length exceeds this value?

[3.92 mm]

CHAPTER 8

Radiation and Antennas

Aims: to give a summary of the theory and ideas behind the design of antennas used for transmitting and receiving electromagnetic radiation.

8.1 SHORT DIPOLE ANTENNA

So far in this book, all the fields have been defined in terms of boundary conditions rather than in terms of the sources of the fields. The one exception is Section 1.6, where the scalar and vector potentials, V and $\mathbf{A}$, were defined in terms of an applied electrical potential difference or electric current flow respectively. However, in Section 1.6, the definitions are confined to static field conditions. In order to calculate the electromagnetic fields radiated from an aerial, expressions similar to eqns. (1.64) and (1.71) are required in terms of time-varying fields. It has been shown in Chapter 4 that time-varying fields are governed by the wave equation and that such fields propagate away from their source. In this section we determine the equations relating time-varying fields to their sources.

Take the definition of the vector potential from eqn. (1.46) and substitute into Faraday's law from Maxwell's equations, eqn. (1.42), to give

$$\nabla \times \mathbf{E} = -\nabla \times \frac{\partial \mathbf{A}}{\partial t} \tag{8.1}$$

E is also defined in terms of the scalar potential by eqn. (1.23) which is consistent with eqn. (8.1) because of the vector identity given in eqn. (1.44), $\nabla \times \nabla V = 0$. Equation (8.1) is a curl equation which may be integrated or uncurled to give

$$\mathbf{E} = -\frac{\partial \mathbf{A}}{\partial t} - \nabla V \tag{8.2}$$

where the constant of integration is provided by eqn. (1.23). Substitute the vector potential into the Maxwell–Ampère law, eqn. (1.52), to give

$$\nabla \times \nabla \times \mathbf{A} = \mu_0 \left(\mathbf{J} + \varepsilon_0 \frac{\partial \mathbf{E}}{\partial t} \right) \tag{8.3}$$

Differentiating eqn. (8.2) with respect to time and substituting for **E** in eqn. (8.3) gives

$$\nabla \times \nabla \times \mathbf{A} = \mu_0 \mathbf{J} - \varepsilon_0 \left(\frac{\partial^2 \mathbf{A}}{\partial t^2} + \nabla V \right) \tag{8.4}$$

Expanding the left-hand side of eqn. (8.4) by substitution from eqn. (1.46) gives

$$\nabla(\nabla \cdot \mathbf{A}) - \nabla^2 \mathbf{A} = \mu_0 \mathbf{J} - \mu_0 \varepsilon_0 \frac{\partial^2 \mathbf{A}}{\partial t^2} - \mu_0 \varepsilon_0 \nabla \frac{\partial V}{\partial t} \tag{8.5}$$

In a static system, the divergence of **A** has been defined to be zero in eqn. (1.67). In a time-varying system it may now be defined by

$$\nabla \cdot \mathbf{A} = -\mu_0 \varepsilon_0 \frac{\partial V}{\partial t} \tag{8.6}$$

which is still consistent with eqn. (1.67) when $\partial/\partial t = 0$. Equation (8.6) is not the only possible definition of **A** but it is the one most commonly used and is called the *Lorentz condition*. Taking the gradient of eqn. (8.6) and substituting into eqn. (8.5) simplifies eqn. (8.5) to

$$\nabla^2 \mathbf{A} - \mu_0 \varepsilon_0 \frac{\partial^2 \mathbf{A}}{\partial t^2} = -\mu_0 \mathbf{J} \tag{8.7}$$

The left-hand side of eqn. (8.7) is similar to eqns. (4.12) and (4.13), the wave equation, so we postulate that eqn. (8.7) describes a field quantity which radiates as a wave out from the source. It has solutions of the form $f(r - ct)$, where r is the radial distance from the source and c is the speed of light.

We already know that electromagnetic waves take a finite time to propagate from the source to the receiver, so we postulate a solution to eqn. (8.7) which uses a value of the source quantity *existing at a time r/c previous to the time in question*. Field values obtained from this relationship are called *retarded potentials*. Apart from the time-varying expression, eqn. (8.7) is similar to eqn. (1.68), whose solution is given in eqn. (1.71). Therefore by similarity, the solution to eqn. (8.7) is

$$\mathbf{A} = \frac{\mu_0}{4\pi} \int_{\text{volume}} \frac{[\mathbf{J}]}{r} dv' \tag{8.8}$$

where [**J**] implies that the value of **J** used is that existing at a time r/c previous to the time in question and, as in eqn. (1.71), the volume of integration is the volume containing the electric currents and not the volume containing the point at which **A** is required. Therefore

$$[\mathbf{J}] = f\!\left(x, y, z, t - \frac{r}{c}\right)$$

226 RADIATION AND ANTENNAS

The value of the retarded potential given by eqn. (8.8) may be used to calculate the values of the magnetic field quantities of the field from eqn. (1.46). However, in order to calculate the electric field quantities, the scalar potential, V, is also required.

Taking Gauss's law, eqn. (1.26), and substituting the material relationship, eqn. (1.1), gives

$$\nabla \cdot \mathbf{E} = \frac{\rho}{\varepsilon_0} \tag{8.9}$$

Taking the divergence of eqn. (8.2) and eliminating $\mathbf{E}$ between eqns. (8.2) and (8.9) gives

$$\frac{\partial}{\partial t} \nabla \cdot \mathbf{A} + \nabla \cdot \nabla V = -\frac{\rho}{\varepsilon_0} \tag{8.10}$$

From eqn. (1.28) it is seen that the second term in eqn. (8.10) is ∇^2. Substitution from eqn. (8.6) is used to eliminate $\mathbf{A}$ from eqn. (8.10). Therefore

$$\nabla^2 V - \mu_0 \varepsilon_0 \frac{\partial^2 V}{\partial t^2} = -\frac{\rho}{\varepsilon_0} \tag{8.11}$$

The left-hand side of eqn. (8.11) is a wave equation similar to eqn. (8.7), so its solution is also a retarded potential. Otherwise, the solution is similar to eqn. (1.64):

$$V = \frac{1}{4\pi\varepsilon_0} \int_{\text{volume}} \frac{[\rho]}{r} dv' \tag{8.12}$$

where $[\rho]$ is the retarded value of the charge density and the volume for the integration is the volume containing the charges and not the volume containing the point at which V is measured, as with eqn. (8.8). Equations (8.8) and (8.12) can now be used to calculate the field radiated from an antenna.

Consider the short dipole antenna shown in Figure 8.1. The antenna is defined to be short in relation to the characteristic wavelength related to the frequency so that the current may be assumed to be constant along the length of the antenna. The current in the antenna is defined as

$$I = I_0 \sin \omega t$$

Because the current is assumed to be constant, the integration in eqn. (8.8) is just a multiplication by the length of the antenna. Then from eqn. (8.8), the vector potential at a distance r from the antenna is given by

$$\mathbf{A} = \frac{\mu_0 \mathbf{I}_0 l \sin \omega(t - r/c)}{4\pi r} \tag{8.13}$$

SHORT DIPOLE ANTENNA

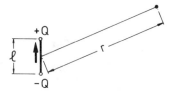

Figure 8.1 A short dipole antenna.

We can substitute $\beta = \omega/c$ from eqn. (4.20) and expand eqn. (8.13) to give

$$\mathbf{A} = \frac{\mu_0 I_0 l}{4\pi r}(\cos\beta r \sin\omega t - \sin\beta r \cos\omega t) \tag{8.14}$$

As with all our earlier expressions, the time variation in eqn. (8.14) will be assumed and the equation written in phasor notation to give

$$\mathbf{A} = \frac{\mu_0 I_0 l}{4\pi r}(\cos\beta r - j\sin\beta r) = \frac{\mu_0 I_0 l}{4\pi r}\exp(-j\beta r) \tag{8.15}$$

The vector potential **A** is always parallel to the current **I** and independent of the angle which r makes with the axis of the dipole.

Using eqn. (1.46), the vector potential may be used to find the magnetic field components of the field radiating from the antenna. Therefore

$$\mathbf{H} = \frac{1}{\mu_0}\nabla \times \mathbf{A} \tag{8.16}$$

However, the antenna is assumed to be at the centre of an otherwise empty space and only the radial distance r enters into eqn. (8.15); therefore it is easiest to operate using a spherical coordinate system, r, θ, ϕ, as shown in Figure 8.2. Also shown in Figure 8.2 are the three components of the vector related to the coordinate directions. In order to apply eqn. (8.16) it is necessary to know the expressions for the curl operation in a spherical coordinate system. Using eqn. (1.46) to provide the notation, they are

$$\left. \begin{array}{l} B_r = \dfrac{1}{r^2 \sin\theta}\left[\dfrac{\partial(r\sin\theta\, A_\phi)}{\partial\theta} - \dfrac{\partial(rA_\theta)}{\partial\phi}\right] \\[8pt] B_\theta = \dfrac{1}{r\sin\theta}\left[\dfrac{\partial A_r}{\partial\phi} - \dfrac{\partial(r\sin\theta\, A_\phi)}{\partial r}\right] \\[8pt] B_\phi = \dfrac{1}{r}\left[\dfrac{\partial(rA_\theta)}{\partial r} - \dfrac{\partial A_r}{\partial\theta}\right] \end{array} \right\} \tag{8.17}$$

For our short dipole antenna, the vector potential **A** resolves into the

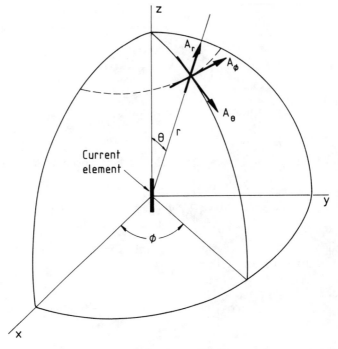

Figure 8.2 Field components in a spherical coordinate system.

following components:

$$A_r = \frac{\mu_0 I_0 l}{4\pi r} \cos\theta \exp(-j\beta r)$$
$$A_\theta = -\frac{\mu_0 I_0 l}{4\pi r} \sin\theta \exp(-j\beta r) \quad \quad (8.18)$$
$$A_\phi = 0$$

For this particular antenna, because of symmetry, $\partial/\partial\phi = 0$. Then substituting into eqn. (8.16) and using eqn. (8.17) gives

$$H_r = 0$$
$$H_\theta = 0$$
$$H_\phi = \frac{I_0 l}{4\pi} \sin\theta \left(\frac{j\beta}{r} + \frac{1}{r^2}\right) \exp(-j\beta r) \quad \quad (8.19)$$

We do not know what charges are needed at the antenna to sustain the current flow, so we cannot use the scalar potential V in conjunction with the vector potential **A** to determine the electric field components of the radiated field. However, Maxwell's equations in a non-conducting medium apply at the point at which the field quantities are to be determined. The magnetic field intensity given by eqn. (8.19) is substituted into eqn. (4.11) to find the components of the electric field intensity. Again using eqn. (8.17),

$$E_r = \frac{I_0 l}{2\pi\varepsilon_0\omega} \sin\theta \left[\frac{\beta}{r^2} - \frac{j}{r^3}\right] \exp(-j\beta r) \tag{8.20}$$

$$E_\theta = \frac{I_0 l}{4\pi\varepsilon_0\omega} \sin\theta \left[\frac{j\beta^2}{r} + \frac{\beta}{r^2} - \frac{j}{r^3}\right] \exp(-j\beta r) \tag{8.21}$$

$$E_\phi = 0$$

The total fields generated by the antenna are given by eqns. (8.19)–(8.21). They have components proportional to $1/r$, $1/r^2$ and $1/r^3$, which are called the radiation field, the induction field and the electrostatic field respectively. The intrinsic impedance from eqn. (4.26) and eqn. (4.20) is used to simplify eqns. (8.20) and (8.21) as an aid to the understanding of these field quantities. Therefore the total fields generated by the antenna are

$$H_\phi = \frac{I_0 l}{4\pi} \sin\theta \left[\frac{j\beta}{r} + \frac{1}{r^2}\right] \exp(-j\beta r) \tag{8.19}$$

$$E_r = \frac{\eta I_0 l}{2\pi} \sin\theta \left[\frac{1}{r^2} - \frac{j}{\beta r^3}\right] \exp(-j\beta r) \tag{8.22}$$

$$E_\theta = \frac{\eta I_0 l}{4\pi} \sin\theta \left[\frac{j\beta}{r} + \frac{1}{r^2} - \frac{j}{\beta r^3}\right] \exp(-j\beta r) \tag{8.23}$$

A further aid to understanding these relationships is obtained by substituting $\beta = 2\pi/\lambda$ from eqn. (3.14). Then it is possible to separate the field quantities into those parts that are significant for small r close to the antenna, the near field, and those parts that are significant for large r, the far field.

The region where the *near field* is significant is called the Fresnel zone. The components of the near field are

$$H_\phi = \frac{I_0 l}{4\pi r^2} \sin\theta \exp(-j\beta r) \tag{8.24}$$

$$E_r = \frac{\eta I_0 l}{2\pi r^2} \sin\theta \left(1 - \frac{j\lambda}{2\pi r}\right) \exp(-j\beta r) \tag{8.25}$$

$$E_\theta = \frac{\eta I_0 l}{4\pi r^2} \sin\theta \left(1 - \frac{j\lambda}{2\pi r}\right) \exp(-j\beta r) \tag{8.26}$$

Close to the antenna, where $r \ll \lambda$, the dominant terms in the expressions for the fields show that the electric and magnetic field quantities are 90° out of phase and these fields do not contribute to the power flow from the antenna. The effect is capacitive and is due to electromagnetic induction and electrostatic fields.

230 RADIATION AND ANTENNAS

The region where the *far field* is significant is called the *Fraunhofer region*. The far field is significant when $r \gg \lambda$. The components of the far field are

$$H_\phi = \frac{jI_0 l}{2\lambda r} \sin\theta \exp(-j\beta r) \tag{8.27}$$

$$E_\theta = \frac{j\eta I_0 l}{2\lambda r} \sin\theta \exp(-j\beta r) \tag{8.28}$$

The wave is propagating radially away from the antenna with a spherical wavefront. At any point on the wavefront where the curvature of the wavefront may be ignored, the electromagnetic radiating wave has all the properties of a plane wave. The electric and magnetic fields lie in the plane of the wavefront perpendicular to the direction of propagation. The electric and magnetic fields are perpendicular to each other, and are in phase. They are related by the intrinsic impedance,

$$E_\theta = \eta H_\phi \tag{8.29}$$

Therefore each elemental part of the spherically propagating wave approximates to a plane wave. The magnitude of the electric field strength in air or free space, where $\eta = 120\pi$, is given by

$$|E| = \frac{60\pi I_0 l \sin\theta}{\lambda r} \tag{8.30}$$

The radiation field is symmetrical about a line along the axis of the antenna, the z-direction in Figure 8.2, and varies sinusoidally in the other plane. If the field strength is plotted as a radial coordinate, it gives a doughnut field pattern as shown in Figure 8.3.

The power flow density in the field is obtained from the Poynting vector from eqn. (1.62). Therefore the average power flow density is given by

$$|S| = \tfrac{1}{2} E_\theta H_\phi = \frac{\eta I_0^2 l^2 \sin^2\theta}{8\lambda^2 r^2}$$

The total power flow is obtained by integrating the expression for the power

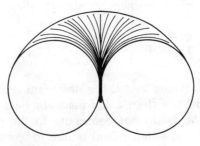

Figure 8.3 The radiation pattern of a short dipole antenna.

SHORT DIPOLE ANTENNA

flow density across the surface of the enclosing sphere as shown in Figure 8.4. Therefore

$$W = \int_0^\pi \int_0^{2\pi} |S| r^2 \sin\theta \, d\theta \, d\phi$$

$$= \frac{\eta I_0^2 l^2}{8\lambda^2} \int_0^\pi \int_0^{2\pi} \sin^3\theta \, d\theta \, d\phi$$

$$= \frac{\pi \eta I_0^2 l^2}{4\lambda^2} \int_0^\pi \sin^3\theta \, d\theta = \frac{\pi \eta I_0^2 l^2}{3\lambda^2} \quad (8.31)$$

The power calculated in eqn. (8.31) flows out into space from the antenna and gives rise to a resistive component of the impedance that the antenna shows to its source. It is called the *radiation resistance* of the antenna. In terms of resistance, the power flow in a circuit is given by

$$W = \tfrac{1}{2} R I_0^2$$

whence, by comparison with eqn. (8.31),

$$R = \frac{2\pi\eta}{3} \left(\frac{l}{\lambda}\right)^2 \quad (8.32)$$

In air, where $\eta = 120\pi = 377\,\Omega$, eqn. (8.32) becomes

$$R = 790 \left(\frac{l}{\lambda}\right)^2 \Omega$$

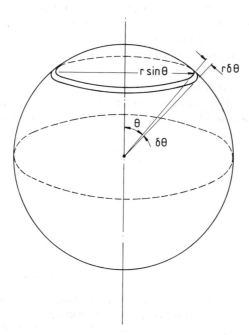

Figure 8.4 Element of integration on the surface of a sphere.

SUMMARY

The electromagnetic field radiated from a time-varying source is determined in terms of *retarded potentials*:

$$\mathbf{A} = \frac{\mu_0}{4\pi} \int_{\text{volume}} \frac{[\mathbf{J}]}{r} dv' \qquad (8.8)$$

where [J] implies that the value of **J** used is that existing at time r/c previous to the time in question. Similarly

$$V = \frac{1}{4\pi\varepsilon_0} \int_{\text{volume}} \frac{[\rho]}{r} dv' \qquad (8.12)$$

The Lorentz condition is

$$\nabla \cdot \mathbf{A} = -\mu_0 \varepsilon_0 \frac{\partial V}{\partial t} \qquad (8.6)$$

$$\mathbf{H} = \frac{1}{\mu_0} \nabla \times \mathbf{A} \qquad (8.16)$$

$$\mathbf{E} = -\frac{\partial \mathbf{A}}{\partial t} - \nabla V \qquad (8.2)$$

For any antenna, the region where the *near field* is significant, where $r \ll \lambda$, is called the *Fresnel zone*. The electric and magnetic fields are out of phase and do not contribute to the power flow from the antenna. Expressions for the field quantities are given in eqns. (8.24)–(8.26).

The region where the *far field* is significant, where $r \gg \lambda$, is called the *Fraunhofer region*. The fields constitute a wave propagating radially away from the antenna with a spherical wavefront. Each elemental part of the spherically propagating wave approximates to a plane wave. Expressions for the field quantities are given in eqns. (8.27) and (8.28). In air

$$|E| = \frac{60\pi I_0 l \sin\theta}{\lambda r} \qquad (8.30)$$

The *radiation resistance* is

$$R = \frac{2\pi\eta}{3} \left(\frac{l}{\lambda}\right)^2 \qquad (8.32)$$

Example 8.1 The *gain* of an antenna is defined as the ratio of the power density radiated in the direction of the maximum radiation to the power density radiated by an isotropic radiator having the same total power. Find the gain of a short dipole antenna.

Answer The total power radiated by the short dipole antenna is given by eqn. (8.31). Therefore the power density radiated by an isotropic antenna having the same total power is that power divided by the surface area of the enclosing sphere, $4\pi r^2$:

$$\text{power density} = \frac{\eta I_0^2 l^2}{12\lambda^2 r^2}$$

The maximum power density radiated by the short dipole antenna is given by

$$\text{power density} = \frac{|E_{max}|^2}{2\eta}$$

Substituting from eqn. (8.30) for the condition when $\sin\theta = 1$ gives

$$\text{power density} = \frac{\eta I_0^2 l^2}{8\lambda^2 r^2}$$

The ratio of these two power density expressions gives the gain. Therefore

$$\text{gain} = 1.5$$

PROBLEM

8.1 A short dipole antenna has a total radiated power of 1.0 kW. Calculate the maximum electric field strength in the radiated field at a distance of 1.0 km from the antenna.

[0.212 V/m]

8.2 ANTENNA ARRAYS

The single short dipole antenna has a radiation pattern which is omnidirectional in the plane perpendicular to the axis of the antenna. Such an antenna is ideal for use in broadcasting where the radiated signal is required to cover the maximum geographical area. However, for other applications the signal may only be required at one specific position and the rest of the radiated power is wasted. Such controlled directional properties may be obtained by combining a number of single antennas into an array. The simplest form of the array is the *linear array*. It may be analysed by a simple addition of the radiated power from each antenna while making allowance for any phase difference in each received signal.

Consider a linear array of N identical individual antennas as shown in Figure 8.5. $n\lambda$ is the physical distance between each antenna, where λ is the characteristic wavelength appropriate to the frequency of the radiated signal. ϕ is the phase difference between any adjacent pair of antennas and is

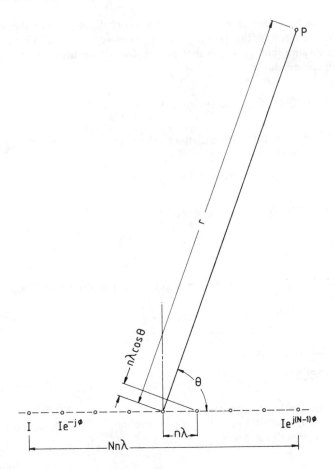

Figure 8.5 A linear array of N antenna elements.

assumed to be uniform along the array. r and θ specify the point at which the radiation is to be measured relative to the centre line of the array. It is assumed that $r \gg Nn\lambda$, the total length of the array, so that any change in the angle θ measured with respect to the ends of the array is negligible. Then the total electric field intensity received at the point P is the sum of the intensities from each antenna:

$$E_\theta = f(r, \theta, I)\{1 + \exp[j(2\pi n \cos \theta - \phi)] + \cdots \\ + \exp[j(N-1)(2\pi n \cos \theta - \phi)]\} \qquad (8.33)$$

where $f(r, \theta, I)$ is the expression for the electric field intensity due to a single antenna in the array. For example, if the array was an array of short dipole antennas, $f(r, \theta, I)$ would be given by the right-hand side of eqn. (8.26). The right-hand side of eqn. (8.33) is a power series which may be summed to give

$$E_\theta = f(r, \theta, I)\left[\frac{\exp[jN(2\pi n \cos \theta - \phi)] - 1}{\exp[j(2\pi n \cos \theta - \phi)] - 1}\right] \qquad (8.34)$$

ANTENNA ARRAYS

The *space factor* of an array is defined as the ratio of the radiation pattern of the array to the radiation pattern of a single element of the array. Therefore the space factor of our linear array is given by

$$S = \frac{\exp[jN(2\pi n \cos\theta - \phi)] - 1}{\exp[j(2\pi n \cos\theta - \phi)] - 1} \tag{8.35}$$

Because the space factor is only a magnitude quantity, any phase difference between the numerator and denominator in eqn. (8.35) may be disregarded and it can be expanded to give

$$S = \frac{\sin\tfrac{1}{2}N(2\pi n \cos\theta - \phi)}{\sin\tfrac{1}{2}(2\pi n \cos\theta - \phi)} \tag{8.36}$$

The expansion into eqn. (8.36) may be understood more easily by considering the phasor diagram shown in Figure 8.6. The electric field intensity due to each element of the array is given by the phasors AB, BC, etc. and the electric field intensity due to the array of antennas is given by AF. Therefore from the diagram, the space factor is given by

$$S = \frac{AF}{AB} = \frac{\sin\tfrac{1}{2}N\alpha}{\sin\tfrac{1}{2}\alpha} \tag{8.37}$$

where $\alpha = 2\pi n \cos\theta - \phi$, so that eqn. (8.37) is the same as eqn. (8.36). If the signal in all the elements of the array are in phase, $\phi = 0$, and eqn. (8.36) simplifies to

$$S = \frac{\sin(\pi N n \cos\theta)}{\sin(\pi n \cos\theta)} \tag{8.38}$$

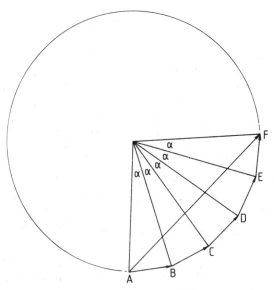

Figure 8.6 Calculation of the space factor of a linear array.

The space factor, S, specifies the manner in which the radiation pattern of a single element is modified by the presence of the additional elements. The total radiation pattern of the array is that of a single element modified by the space factor. In a simple array, the dipole pattern specifies the field strength in a direction including the plane of the dipole, whereas the array specifies the field strength variation in the plane perpendicular to the dipole. From eqn. (8.36), the maximum field strength occurs when

$$\cos \theta = \frac{\phi}{2\pi n} \qquad (8.39)$$

Expanding eqn. (8.37) as $\alpha \to 0$ gives

$$S = \frac{\sin \frac{1}{2} N \alpha}{\sin \frac{1}{2} \alpha} = \frac{\frac{1}{2} N \alpha}{\frac{1}{2} \alpha} = N \qquad (8.40)$$

Then the maximum field strength due to a multi-element array is equal to N times the field strength due to a single element. When the driving signal to all the elements is in phase, $\phi = 0$ and the maximum field occurs perpendicular to the line of the array, where $\theta = \pi/2$.

The in-phase array with the maximum field strength perpendicular to the line of the array is called a *broadside array*. From eqn. (8.38) it is seen that there will be zeros in the field strength occurring on bearings given by

$$\pi N n \cos \theta = \pi, 2\pi, 3\pi, \ldots$$

Then

$$\cos \theta = \frac{1}{Nn}, \frac{2}{Nn}, \frac{3}{Nn}, \ldots$$

There is a subsidiary maximum in the field strength between each of these minima. The position of the maximum field strength of an antenna or an antenna array is called the *main beam* of the antenna and the subsidiary maxima are called *side lobes*. Usually antenna are designed to minimise the electromagnetic power lost in side lobes. Sometimes minimum side lobes are a design requirement so as to minimise interference from unwanted radiated signals. The radiation pattern for a two-element broadside array is shown in Figure 8.7.

Usually designers and users of antennas need to know the width of the main beam, that is the angular distance occupied by the main beam. The simplest definition would appear to be the angle between the zeros each side of the main beam. However, this may often be impractical as a general definition because many antennas do not have any directions of zero energy. The definition chosen is similar to that involved in the calculation of the Q-factor of a resonant circuit, as shown in Figure 6.15; it is the angle between those points on the main beam where the energy has dropped to

ANTENNA ARRAYS 237

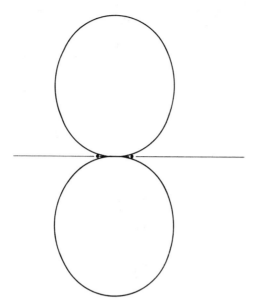

Figure 8.7 The radiation pattern for a broadside array with two elements spaced half a wavelength apart.

half its peak value at the point of the beam. Such an angle is called the *beamwidth* of the antenna.

For a broadside array, the beamwidth is obtained from eqn. (8.38). When

$$\frac{\sin(\pi N n \cos \theta_1)}{\sin(\pi n \cos \theta_1)} = \frac{N}{\sqrt{2}} \qquad (8.41)$$

the beamwidth $= 2(\pi/2 - \theta_1)$. For an array with a large number of elements, $N \gg 1$, and we get the approximate result that

$$\cos \theta_1 \approx \frac{1}{2Nn}$$

Because, for a broadside array, the main beam is at $\theta = 90°$, the beamwidth of the array is given by

$$BW = 2\sin^{-1}\left(\frac{1}{2Nn}\right) \qquad (8.42)$$

The radiation pattern for a ten-element broadside array is shown in Figure 8.8.

The beamwidth of an array is inversely proportional to the number of elements in the array. The number of sidelobes and the number of zeros is directly proportional to the number of elements in the array. The larger the array, the larger the number of sidelobes. The theory of a broadside array

Figure 8.8 The radiation pattern for a broadside array with ten elements spaced half a wavelength apart.

assumes that the currents in each element of the array are of equal amplitude and phase. A large array can tolerate quite large variations in the amplitude and phase of the currents in each element without making a significant change to the direction of the main beam. The most obvious effect of any variation from the ideal is a loss of efficiency, i.e. a reduction in the power of the main beam and an increase in the power directed into the sidelobes and the filling in of the zeros between the sidelobes. Therefore large arrays are used to provide signals for long-distance communication, because it does not matter whether they are very accurately driven. Alternatively, the direction of the main beam of a short array having only a few elements may be controlled by varying the phase of the currents driving the elements of the array. Controlled phase variation is used to guide the direction of the main beam in electronically scanned radar aerials. Controlled non-uniform current distribution to each element of the array may be used to reduce the size of the sidelobes.

An array can also be used to provide a radiated signal along the line of the array. Such an array is called an *endfire array*. It has the maximum radiated signal at $\theta = 0$. Then it is seen that the maximum in S from eqn. (8.36) for

$\theta_1 = 0$ occurs when

$$\phi = 2\pi n \qquad (8.43)$$

The phase difference between the elements along the array is proportional to a travelling wave pattern and the array may be fed automatically from a line propagating in the TEM-mode. Substituting from eqn. (8.43) into eqn. (8.36) shows that the zeros of field strength in the radiation pattern of an endfire array occur when

$$\tfrac{1}{2}N(2\pi n \cos\theta - 2\pi n) = k\pi$$

where k is any integer. Therefore

$$\cos\theta - 1 = \frac{k}{Nn}$$

and, as with a broadside array, the maxima of the sidelobes occur halfway between these zeros. The space factor for an endfire array is given by

$$S = \frac{\sin \pi Nn(\cos\theta - 1)}{\sin \pi n(\cos\theta - 1)} \qquad (8.44)$$

The radiation pattern for a two-element endfire array is shown in Figure 8.9. The essential characteristics of an endfire array may be seen from the figure. The main beam points along the length of the array and there is a zero of field in the opposite direction.

An array need not be confined to one dimension. It is possible to have an array of arrays and to have a two-dimensional planar array. It is possible to combine a number of broadside arrays in parallel to make an endfire array, so eliminating one of the main beams of the broadside array and eliminating a major proportion of the unwanted radiation. If two ten-element broadside arrays are arranged a quarter-wavelength apart as an endfire array, the space factor of the broadside array shown in Figure 8.8 is multiplied by the

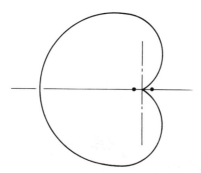

Figure 8.9 The radiation pattern for an endfire array with two elements spaced a quarter wavelength apart.

space factor of the two-element endfire array shown in Figure 8.9 to give the combined radiation pattern shown in Figure 8.10. In all of the arrays discussed so far, the array determines the radiation pattern in the horizontal plane and the radiation pattern in a vertical plane is determined by that of the individual antenna elements forming the array. At lower frequencies even the short dipole antenna is large and the individual antenna elements are the only way to control the vertical radiation pattern. At higher frequencies, dipoles are smaller and a number of dipoles may be mounted vertically to provide a broadside array in the vertical plane. Then all possible combinations of arrays in three dimensions may be used to control the radiation pattern of an aerial.

At very high frequencies, the short dipole antenna becomes small and a large number of individual antennas may be mounted in a plane to give a two-dimensional array. In a radar system, control of the amplitude and phase of the signal in each individual antenna in such a planar array is used to sweep the radiated beam around the sky and also to control the nulls between the sidelobes so as to eliminate reception of interfering signals. One such radar system has the antennas mounted on the sloping faces of a triangular building with about 5000 individual antennas on each face.

The carefully controlled amplitude and phase of the signal applied to each element of an array are acceptable when the complexity and expense is justified by the particular application. However, for mass-produced consumer applications such expense cannot be justified. However, a system has been devised where parasitic elements can be used to form an endfire array. A parasitic element is not connected to any driver current. It is placed near to the driven antenna element in the array so that a current is induced in the parasitic element by mutual field coupling. The parasitic element then acts as an independent but weaker radiator and the resultant radiation pattern is a combination of the fields due to both the driven and the parasitic

Figure 8.10 The radiation pattern of a two-element endfire array of two ten-element broadside arrays, being a combination of the radiation patterns shown in Figures 8.8 and 8.9.

antennas. If the elements in the array are near a resonant length, the phase of the currents in the parasitic elements may be controlled by altering their length, since the reactance of a dipole changes from capacitive to inductive as the length of the dipole is tuned through resonance.

Typical radiation patterns for a driven antenna with one parasitic element of different lengths are shown in Figure 8.11. The elements are closely spaced at about 0.1λ apart. The use of a resonant length parasitic element increases the radiated power along the length of the array by a factor of two but it is equal in two opposite directions. The use of a longer parasitic element as shown in Figure 8.11(b) directs the main beam away from the parasitic element and the parasitic element is called a *reflector*. The use of a shorter parasitic element as shown in Figure 8.11(c) directs the main beam in the same direction as the parasitic element and in this case the parasitic element is called a *director*. Both a reflector and a number of directors may be combined with a driven dipole to give a greater directional gain. Such an aerial is shown in Figure 8.12. Aerials of this type are called *Yagi arrays* and are commonly used for television reception.

The astute reader may well be wondering what justification there is for jumping from a description of driven antennas generating a radiated field at the beginning of this section to television receiving antennas in the last paragraph. Does a receiving antenna work in exactly the same way as a transmitting antenna? The answer is yes. It is an example of the application of the *reciprocity principle*. Formally, the reciprocity principle may be stated as: for any passive circuit, if a current I injected into the input terminals gives rise to a voltage V appearing across the open-circuit output terminals, the same current I injected into the output terminals will give rise to the same voltage V appearing across the open circuit input terminals. Equally, the reciprocity principle may be applied to antennas. The radiation pattern

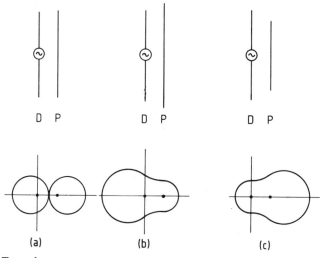

Figure 8.11 Two-element parasitic antenna arrays: (a) resonant parasitic element; (b) long parasitic element is a reflector; (c) short parasitic element is a director.

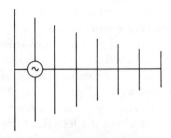

Figure 8.12 Yagi antenna array as used as a television receiving aerial.

of a driven antenna array is the same as that of the same array used as a receiving aerial. Any pair of aerials which are situated so as to be within the radiation pattern of the other constitute a passive circuit as far as the reciprocity principle is concerned. Therefore the radiation pattern, whether calculated or measured, of any antenna or array is the same whether the antenna is being used as a transmitter or a receiver.

SUMMARY

The *space factor* of an array

$$S = \frac{\sin \tfrac{1}{2} N(2\pi n \cos \theta - \phi)}{\sin \tfrac{1}{2}(2\pi n \cos \theta - \phi)} \tag{8.35}$$

The space factor of a *broadside array*

$$S = \frac{\sin (\pi N n \cos \theta)}{\sin (\pi n \cos \theta)} \tag{8.38}$$

Zeros occur in the radiated field pattern at

$$\cos \theta = \frac{1}{Nn}, \frac{2}{Nn}, \frac{3}{Nn}, \ldots$$

with subsidiary maxima between them which are called *sidelobes*.
    The beamwidth is given by

$$\frac{\sin (\pi N n \cos \theta)}{\sin (\pi n \cos \theta)} = \frac{N}{\sqrt{2}} \tag{8.41}$$

For a large array where $N \gg 1$, the beamwidth is

$$BW = 2 \sin^{-1}\left(\frac{1}{2Nn}\right) \tag{8.42}$$

For an *endfire array*,

$$S = \frac{\sin \pi N n(\cos \theta - 1)}{\sin \pi n(\cos \theta - 1)} \quad (8.44)$$

Parasitic antennas are used in the *Yagi array*. A parasitic longer than resonant length is a *reflector* and a parasitic shorter than resonant length is a *director*.

The reciprocity principle may be applied to antennas so that the radiation pattern of an antenna used as a transmitting aerial is the same as that of the same antenna used as a receiving aerial.

Example 8.2 Calculate the beamwidth of a ten-element broadside array with $\frac{1}{2}\lambda$ spacing between elements.

Answer A ten-element array is a large array with $N \gg 1$ so that eqn. (8.42) gives the answer to the question. Putting numbers into eqn. (8.42) gives

$$BW = 2\sin^{-1} 0.10 = 11.5°$$

This beamwidth may be observed by inspection of Figure 8.8.

Example 8.3 Show that the size of the first side lobe of a large broadside array is 21% of the size of the main beam.

Answer The size of the maximum of the main beam of an array is N times the size of the radiation from a single element of the array. The first sidelobe maximum is assumed to be located half-way between the first two nulls of the radiation pattern. The nulls are situated at angles given by

$$\cos \theta = \frac{1}{Nn}, \frac{2}{Nn}$$

Then the maximum of the side lobe is situated at

$$\cos \theta = \frac{3}{2Nn}$$

Then from eqn. (8.38), the space factor gives the amplitude at that angle to be

$$S = \frac{\sin(3\pi/2)}{\sin(3\pi/2N)} \approx \frac{1}{(3\pi/2N)} = \frac{2N}{3\pi} = 0.21N$$

Since the maximum of the main beam is N relative to the same individual antenna pattern, the amplitude of the sidelobe is 21% of the amplitude of the main beam.

PROBLEMS

8.2 Calculate the beamwidth of a broadside array with a $\frac{1}{2}\lambda$ spacing between elements when the total length of the array is 8λ.

[7.2°]

8.3 A broadside array consists of a number of antenna elements spaced $\frac{1}{2}\lambda$ apart. By considering the sum of a number of two-element arrays, show that the radiation pattern of an array, where the magnitude of the currents in each element follows the binomial coefficients (1, 2, 1; 1, 3, 3, 1; 1, 4, 6, 4, 1 etc.), has no sidelobes.

$[S = 2^{(N-1)} \cos^{(N-1)}(\frac{1}{2}\pi \cos \theta)]$

8.3 LONG ANTENNAS

The short antenna with a uniform current distribution is the simplest to analyse analytically but except at the longest wavelength radiation it is not necessarily the most practical. A resonant length of wire used as an antenna is simpler to drive electrically. The method of analysis is similar to that used in the analysis of antenna arrays. It is assumed that the radiation pattern is measured in the far field, where the radial distance r is much greater than the wavelength and much greater than the total length of the antenna. The geometry of the problem is shown in Figure 8.13. It is assumed that there is

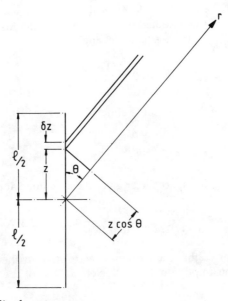

Figure 8.13 A long dipole antenna.

a sinusoidal distribution of current along the length of the antenna:

$$I(z) = I_0 \cos \beta z \qquad (8.45)$$

Then the field at the distant point, (r, θ), due to the small element of the antenna is given by eqn. (8.28):

$$\delta E_\theta = \frac{j\eta I(z)\delta z}{2r\lambda} \sin \theta \exp(-j\beta r) \qquad (8.46)$$

Substituting from eqn. (8.45) for $I(z)$ and integrating along the length of the antenna gives

$$E_\theta = \frac{j\eta}{2r\lambda} \sin \theta \int_{-1/2l}^{1/2l} I_0 \cos \beta z \exp(-j\beta z \cos \theta) \, dz \exp(-j\beta r) \qquad (8.47)$$

Use can be made of the standard integral:

$$\int \exp(az) \cos bz \, dz = \frac{\exp(az)}{a^2 + b^2} (a \cos bz + b \sin bz)$$

If the antenna is a resonant length,

$$l = \frac{n\lambda}{2}$$

where n is the length of the antenna in half-wavelengths. The limits of integration in eqn. (8.47) become $\pm \tfrac{1}{4} n\lambda$ and substituting from eqn. (4.29) for β, at the integral limits,

$$\beta z = \pm \frac{2\pi}{\lambda} \frac{n\lambda}{4} = \pm \tfrac{1}{2} n\pi$$

Then using the standard integral, eqn. (8.47) becomes

$$E_\theta = \frac{j\eta I_0}{2\pi r} \left(\frac{\cos(\tfrac{1}{2} n\pi \cos \theta) - \cos \tfrac{1}{2} n\pi}{\sin \theta} \right) \qquad (8.48)$$

For $n = 1$, the half-wavelength antenna, the radiation pattern is little different from that of the short dipole antenna shown in Figure 8.3. Some radiation patterns for longer resonant antennas are shown in Figure 8.14. As the antenna length is increased, the main beam becomes sharper and the beamwidth is reduced. The beamwidth of a half-wave dipole ($n = 1$) is 78°, whereas that of a one-wavelength dipole ($n = 2$) is only 47°. Beyond a length of about 1.25λ, the main beam starts to reduce and sidelobes appear. When the antenna becomes 2λ long, there is now no radiated field perpendicular to the axis of the antenna and the radiation pattern consists of two major lobes. For even longer lengths of antenna, the radiation pattern consists of a larger

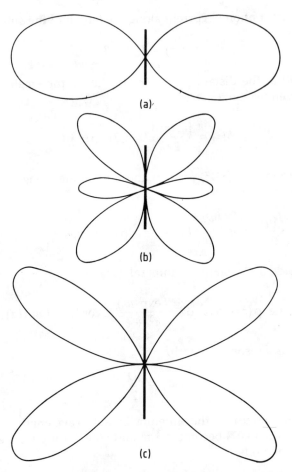

Figure 8.14 Radiation patterns for some long dipole antennas: (a) $l = \lambda$; (b) $l = 1.5\lambda$; (c) $l = 2\lambda$.

number of lobes, with no main beam perpendicular to the axis of the antenna.

A length of wire used as a transmission line also acts as an antenna. This is a travelling wave antenna which is designed to be a large number of wavelengths long and terminated in its characteristic impedance as shown in Figure 8.15. Its space factor may be calculated similarly to our previous calculations as the integrated sum of a large number of individual dipoles. The current in the wire is constant in amplitude provided that the attenuation is zero. The phase of the current changes in accordance with the wavelength of the wave transmitted along the line. If the attenuation is zero, the radiation pattern depends on the length of the antenna. Then by an analysis similar to that used to obtain eqn. (8.48) the field of the antenna is given by

$$E_\theta = \frac{j\eta I_0}{2\pi r} \frac{\sin\theta}{(1-\cos\theta)} \sin\left[\frac{\pi l}{\lambda}(1-\cos\theta)\right] \tag{8.49}$$

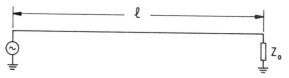

Figure 8.15 Travelling wave antenna.

There are zeros in the radiated field at

$$\sin \theta = 0$$

or

$$\theta = 0, \pi, 2\pi$$

and at

$$\sin \frac{\pi l}{\lambda}(1 - \cos \theta) = 0$$

or

$$\cos \theta = 1, 1 - \frac{\lambda}{l}, 1 - \frac{2\lambda}{l}, \ldots$$

There is a primary maximum situated approximately half-way between the first two zeros at an angle given by

$$\cos \theta = 1 - 0.37\frac{\lambda}{l} \qquad (8.50)$$

and this gives the travelling wave antenna two main beams situated at this angle each side of the line of the antenna wire. The pattern for a long line is shown in Figure 8.16. The pattern is biased in the direction of travel of the wave but it has a null in the true forward direction. The major lobe makes almost the same angle with the axis of the wire as a resonant antenna of the same length. The resonant system is seen to be the vector sum of two equal and opposite travelling waves on the antenna wire.

Four transmission-line antennas may be combined to provide a main beam along the axis of the antenna. They are mounted approximately at the angle of the major lobe to the axis of the antenna, as shown in Figure 8.17, so that four major lobes combine to give a large radiation field along the axis of the antenna with much of the rest of the radiation from each individual antenna wire cancelling out. The angle is chosen so that the main-beam contributions from each wire combine in phase for maximum gain.

Such a combination of transmission wires into an antenna is called a *rhombic antenna*. Losses in practical transmission line antennas result in filling the nulls of the radiation pattern and changing the symmetrical shape of the main lobe.

248 RADIATION AND ANTENNAS

Figure 8.16 Radiation pattern of a travelling wave antenna.

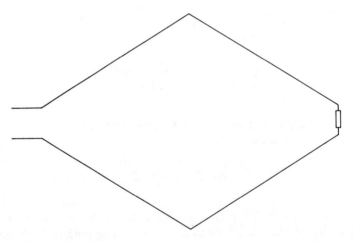

Figure 8.17 Rhombic antenna.

Any antenna adjacent or near to a plane conducting surface induces an image antenna in the conductor. At low frequencies, the wavelength is large and antennas have to be mounted near to the ground. In spite of the relatively poor conductivity of soil compared to that of metal, it is often satisfactory to consider the ground as a perfectly conducting surface. Then low-frequency antennas create electrical images in the ground. The image of a half-wave antenna some distance above the ground is shown in Figure 8.18(a). If a vertical dipole is some distance above the ground, its radiation pattern is unaffected by the presence of the image. However, if the antenna is close to the ground, the image effectively doubles the length of the antenna as shown in Figures 8.18(b) and (c). The currents in the antenna and its image in the ground are in the same direction, the currents are in phase. In the long- and medium-wave band broadcasting antennas use the

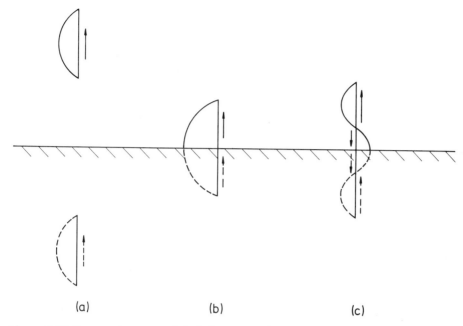

Figure 8.18 Some antennas and their images in the presence of ground showing the current distribution in each antenna: (a) half-wavelength antenna above the ground; (b) quarter-wavelength antenna adjacent to the ground; (c) three-quarter-wavelength antenna adjacent to the ground.

full height of the tower as the antenna. For example, for broadcasting at 750 kHz ($\lambda = 400$ m) a quarter-wavelength tower of height 100 m is equivalent to a half-wave antenna because of the image effect of the ground.

The short dipole antenna can be considered as a lumped circuit consisting of a resistance and a capacitance in series. It is useful to be able to eliminate the capacitance to increase the efficiency of the antenna. It may be tuned using an inductor but the wire of the inductor introduces undesirable resistive loss. An alternative is to increase the effective length of the antenna by adding an additional element to the top of the antenna. This is called *top loading*. The additional element can take the form of a T or inverted L construction as shown in Figure 8.19. As shown in the figure, the currents in the vertical part of the antenna and its image are in phase whereas the currents in the horizontal part of the antenna and its image are out of phase. The radiation from the horizontal part of the antenna is cancelled by its image so it is described as having a non-radiating load. Top loading is a good way of ensuring that the current in the vertical part of the antenna is substantially constant.

At higher frequencies the dipole antenna is both simple and convenient. Wavelengths vary between 10 m and 10 cm. The dipole is mounted at the top of a mast or tower and ground effects are negligible. For example, television broadcasting operates in this band. Television transmitter aerials need to be as omnidirectional as possible in the horizontal plane and to give

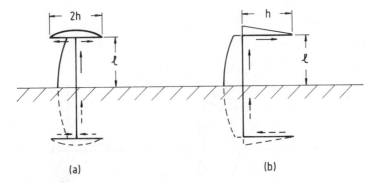

Figure 8.19 Top-loaded antennas and their images adjacent to the ground: (a) T, (b) inverted L, for both $l + h = \lambda/4$.

good ground cover with maximum range. Because of the reflection properties of any support mast, television transmitters use arrays of dipoles around the top portion of the support mast. The dipoles are vertical or horizontal depending on the polarisation required. The vertical array pattern is used to tilt the radiation pattern towards the ground at the maximum range. Television receiving antennas are almost always Yagi arrays as described in Section 8.2.

SUMMARY

For a resonant antenna,

$$E_\theta = \frac{j\eta I_0}{2\pi r} \left(\frac{\cos(\tfrac{1}{2}n\pi \cos\theta) - \cos\tfrac{1}{2}n\pi}{\sin\theta} \right) \tag{8.48}$$

For the non-resonant antenna,

$$E_\theta = \frac{j\eta I_0}{2\pi r} \frac{\sin\theta}{(1 - \cos\theta)} \sin\left[\frac{\pi l}{\lambda}(1 - \cos\theta)\right] \tag{8.49}$$

which has main beams situated at an angle given by

$$\cos\theta = 1 - 0.37\frac{\lambda}{l} \tag{8.50}$$

each side of the line of the antenna wire.

Four transmission-line antennas may be combined to form the *rhombic antenna*, having one major forward beam to the radiation pattern.

The efficiency of a short antenna may be improved by *top loading*, so making it a resonant length.

LONG ANTENNAS 251

Example 8.4 For the same peak current in each antenna wire, compare the maximum radiated far electric field strength in (a) a dipole of length 0.10λ, (b) a resonant wire of length $\frac{1}{2}\lambda$, and (c) a resonant wire of length λ.

Answer In all three antennas, the maximum of the radiated field pattern occurs perpendicular to the axis of the antenna, i.e. when $\theta = \frac{1}{2}\pi$. The field of the short antenna is given by eqn. (8.30) and that of the resonant antennas by eqn. (8.48). Therefore the field strengths of the three antennas are given by

(a) $E_\theta = \dfrac{\eta I_0 l}{2\lambda r} = 0.10 \dfrac{\eta I_0}{2r}$

(b) $E_\theta = \dfrac{\eta I_0}{2\pi r} \dfrac{\cos 0 - \cos \frac{1}{2}\pi}{1} = \dfrac{1}{\pi} \dfrac{\eta I_0}{2r}$

(c) $E_\theta = \dfrac{\eta I_0}{2\pi r} \dfrac{\cos 0 - \cos \pi}{1} = \dfrac{2}{\pi} \dfrac{\eta I_0}{2r}$.

Therefore, the required ratio of field strengths is $0.10: 0.32: 0.64$.

Example 8.5 A receiving antenna consists of a horizontal wire 100 m long. It is terminated with its characteristic impedance at each end and one end is connected to the receiver. It is arranged at 28° with respect to the bearing of the transmitter and it is used to receive two signals, one at 10 MHz and another at 15 MHz. Given that the two transmitted signals are of equal strength, find the ratio of the received signal amplitudes.

Answer The reciprocity principle means that the receiving antenna can be analysed as if it were a transmitting antenna. Then it is a travelling-wave antenna and the field strength is given by eqn. (8.49). It is assumed that the transmitted fields are such that the electric field vector at the antenna is the same for both frequencies of transmission. Then the only contribution to the strength of the received signal is due to the characteristics of the receiving antenna, which are given by eqn. (8.49).

(a) For $f = 10$ MHz:

$$\lambda = \frac{c}{f} = \frac{3 \times 10^8}{10 \times 10^6} = 30 \text{ m}$$

and

$$\frac{l}{\lambda} = \frac{100}{30} = 3.33$$

$$\sin 28° = 0.4695$$

$$1 - \cos 28° = 1 - 0.88295 = 0.11705$$

$$\frac{\pi l}{\lambda}(1 - \cos 28°) = 3.333\pi \times 0.11705 = 70.2°$$

$$\text{Relative signal strength} = \frac{\sin 28° \sin 70.2°}{1 - \cos 28°} = 3.77$$

(b) For $f = 15$ MHz:

$$\lambda = \frac{c}{f} = \frac{3 \times 10^8}{15 \times 10^6} = 20 \text{ m}$$

$$\frac{l}{\lambda} = \frac{100}{20} = 5$$

$$\frac{\pi l}{\lambda}(1 - \cos 28°) = 5\pi \times 0.11705 = 105.3°$$

$$\text{Relative signal strength} = \frac{\sin 28° \sin 105.3°}{1 - \cos 28°} = 3.88$$

Therefore the relative signal strengths are 1.03:1. The travelling wave antenna is a good broadband antenna.

PROBLEMS

8.4 By considering the image charges which support the currents in an antenna and its image, show that the radiation from a vertical antenna close to the ground is strengthened, whereas that from a horizontal antenna close to the ground is weakened.

8.5 Calculate the angles subtended at the axis of the antenna for the zeros in the radiation pattern of a $3\lambda/2$ dipole antenna.

[0°, 48.2°, 70.5°]

8.4 APERTURE ANTENNAS

At higher frequencies, it is often easier to manipulate any modulated wave as electromagnetic radiation rather than currents in wires. The source for a high-frequency antenna becomes an open-ended waveguide or horn. Then the radiated electric field is directly related to the electric field in the aperture of the source. The calculation is similar to that used to calculate the field of an antenna array in section 8.2, except that, in this case, the source field is continuous across the aperture. The geometry of the problem is shown in Figure 8.20. If the field at the distant point $P(r, \theta)$ due to the small element of field $E(x)$ in the aperture is $A(x)$, the total field at P allowing for the phase delay due to the different elements of field is given by

$$E_\theta = \int_{-1/2a}^{1/2a} A(x) \exp\left(-j \frac{\pi x}{\lambda} \cos \theta\right) dx \tag{8.51}$$

The variation of $A(x)$ with x is the same as the variation of $E(x)$ with x. $E(x)$ exists only across the aperture, so that the limits of the integral in eqn. (8.51)

APERTURE ANTENNAS 253

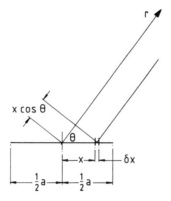

Figure 8.20 An aperture antenna.

may be changed to ±∞. Then eqn. (8.51) becomes a Fourier transform equation

$$E_\theta = \int_{-\infty}^{\infty} A(x) \exp(-jkx)\,dx \qquad (8.52)$$

where $k = (\pi/\lambda)\cos\theta$. If $A(x)$ represents a uniform distribution of the field across the aperture, the solution of eqn. (8.52) is the well-known Fourier transform result,

$$E_\theta = \frac{A \sin \tfrac{1}{2}ak}{\tfrac{1}{2}ak} \qquad (8.53)$$

The same result for the radiation pattern from a uniformly illuminated aperture in eqn. (8.53) may also be obtained from the expression for the uniformly excited broadside array given in eqn. (8.38). The uniformly illuminated aperture is the limit of a very large number of array elements closely spaced together. Then N becomes very large, n becomes infinitesimally small, $Nn\lambda = a$ and the output from each element is proportional to $1/N$. Substituting into eqn. (8.38) gives eqn. (8.53).

The radiation pattern for the uniformly illuminated aperture is given in Figure 8.21. A particular aperture distribution which minimises the amplitude of the sidelobes although the beamwidth is larger is a $\cos^2(\pi x/a)$ distribution. The Fourier transform of $A \cos^2 ka$ is

$$E_\theta = \frac{A \sin \tfrac{1}{2}ka}{\tfrac{1}{2}ka(\pi^2 - \tfrac{1}{4}k^2 a^2)} \qquad (8.54)$$

where $k = (\pi/\lambda)\cos\theta$ as before. The radiation pattern of the $\cos^2$ aperture distribution is also shown in Figure 8.21. The field across the face of any aperture exists in two dimensions. The radiation pattern in any one plane is determined by the variation of the field strength across the line of the

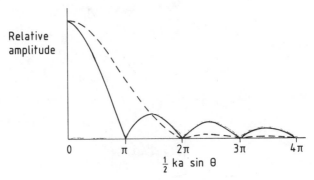

Figure 8.21 Radiation patterns for an aperture antenna: ———— uniform illumination; −−−−−− $\cos^2(\pi x/a)$ illumination.

aperture in that plane. Therefore the radiation pattern of a uniformly illuminated rectangular aperture as shown in Figure 8.22 is given by

$$E_\theta = \frac{A \sin[(\pi a/\lambda)\sin\theta]\sin[(\pi b/\lambda)\sin\phi]}{(\pi a/\lambda)(\pi b/\lambda)\sin\theta\sin\phi} \qquad (8.55)$$

For the open-ended waveguide propagating the dominant TE_{10}-mode, the radiation pattern is the product of that given by the uniform illumination in the vertical plane and that given by a $\cos(\pi x/a)$ distribution in the horizontal plane.

The gain of an antenna is assumed to refer to the direction of maximum radiation intensity since this is usually the interest of any system designer.

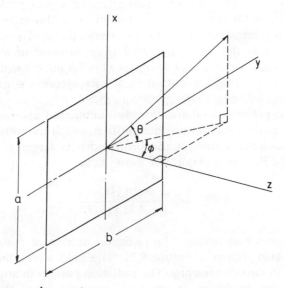

Figure 8.22 A rectangular aperture.

There are two different definitions of antenna gain, the *directive gain* or the directivity and the *power gain* or simply the gain. The directive gain is defined as

$$G_D = \frac{\text{maximum radiation field intensity}}{\text{average radiation field intensity}} \qquad (8.56)$$

The directive gain is found from the radiation pattern. The power gain is defined as

$$G = \frac{\text{power density in the direction of maximum intensity}}{\text{power density from isotropic antenna with the same power input}} \qquad (8.57)$$

The gain of the short dipole antenna was found to be 1.5 in the answer to Example 8.1. The gain of an aperture antenna is calculated similarly, but does involve a double integration to take account of both planes. The uniformly illuminated aperture provides a simple result. For a large aperture producing a relatively narrow beam, the gain of a rectangular aperture as shown in Figure 8.22 is given by

$$G = \frac{4\pi ab}{\lambda^2} \qquad (8.58)$$

and for the more general case of the uniformly illuminated aperture

$$G = \frac{4\pi(\text{aperture area})}{\lambda^2} \qquad (8.59)$$

For other distributions of the electromagnetic fields across the aperture, the gain is less than that given by eqn. (8.59). The effective area is less than the actual area and an efficiency factor k is introduced to allow for the effect. Then the general expression for the gain of any aperture antenna is

$$G = \frac{4\pi kA}{\lambda^2} \qquad (8.60)$$

where k is the efficiency factor and A is the area of the aperture of the antenna. Some factors for some common distributions are given in Table 8.1. As an approximate rule of thumb result, the beamwidth is given by

$$\text{beamwidth} \approx \frac{\lambda}{ka} \text{ radians} \qquad (8.61)$$

Table 8.1 Properties of various antenna aperture distributions.

Type of distribution	Beamwidth (degrees)	First sidelobe intensity	Efficiency factor (k)
uniform $f(x) = 1$	$50.3 \dfrac{\lambda}{a}$	0.217	1.0
cosine $f(x) = \cos(\pi x/2)$	$68.8 \dfrac{\lambda}{a}$	0.071	0.81
triangular $f(x) = 1 - |x|$	$73.4 \dfrac{\lambda}{a}$	0.048	0.75
raised cosine $f(x) = \cos^2(\pi x/2)$	$83.2 \dfrac{\lambda}{a}$	0.027	0.667

A narrow-beamwidth high-gain antenna needs to have a large aperture as shown by eqns. (8.59) and (8.61). However, the cross-sectional dimensions of a waveguide operating in the dominant mode are of the order of half a wavelength or less. The open-ended waveguide is also an inefficient radiator of electromagnetic energy because of the impedance mismatch at its mouth. The open-ended waveguide has a different radiation pattern in its two principal planes because of the different field distributions and the different dimensions in the two perpendicular directions. For the conventional rectangular waveguide shown in Figure 6.1, there is a cosine distribution of field in the horizontal plane, and the horizontal radiation pattern is characterised by a beamwidth of $68.8\lambda/a$ and an efficiency factor $k = 0.81$, whereas there is a uniform distribution of field in the vertical plane and the radiation pattern has a beamwidth of $50.3\lambda/b$ and the efficiency factor is unity. A flare at the end of the waveguide both increases the cross-sectional area of the antenna aperture and also provides an impedance match between the waveguide and the intrinsic impedance of free space. Such horns are frequently used to terminate waveguide feeds to antennas.

Horns flared in just one plane are called sectoral horns and those flared in both planes are called pyramidal horns. Various horns are illustrated in Figure 8.23. The radiation patterns from a horn depend on the flare angle and the length of the horn. Too sharp a flare angle does little to improve the impedance mismatch and also introduces an unacceptably large phase error across the aperture. Any horn introduces some phase error across the aperture due to the circular wavefront generated by the horn as shown in Figure 8.24. A small phase error has little effect on the radiation pattern but a larger phase error defocuses the radiation pattern filling in the sidelobe nulls and reducing the on-axis gain. An acceptable limit for the phase error can be set by $\delta < \lambda/8$ as shown by Figure 8.24. By geometry, it can be shown

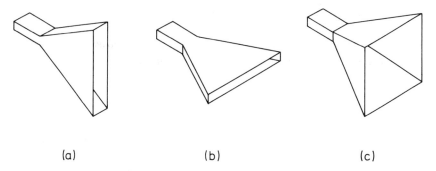

Figure 8.23 Waveguide horns: (a) E-plane; (b) H-plane; (c) pyramidal.

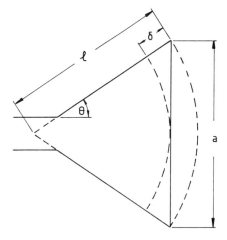

Figure 8.24 Effect of flare angle on a waveguide horn.

that

$$\delta_{max} = \frac{\lambda}{8} \tag{8.62}$$

$$l_{min} = \frac{a^2}{\lambda} + \frac{\lambda}{16} \tag{8.63}$$

$$\sin \theta_{max} = \frac{a}{2l} \approx \frac{\lambda}{2a} \tag{8.64}$$

For many applications, a small aperture is acceptable and these equations may be satisfied by convenient values of l and θ. For large apertures, a horn satisfying these conditions is long and bulky but such horns are in use. However, the waveguide horn has only limited use as a primary antenna. Higher gain and narrower beamwidths are obtained using the reflector antennas described in the next section.

SUMMARY

The radiation pattern of an aperture antenna is the *Fourier transform* of the field distribution in the aperture. For the *uniformly illuminated aperture*,

$$E_\theta = \frac{A \sin\left[(\pi a/2\lambda)\cos\theta\right]}{(\pi a/2\lambda)\cos\theta} \tag{8.53}$$

The $\cos^2$ distribution in the aperture gives the smallest sidelobes.

For a rectangular aperture, the radiation patterns in the two perpendicular planes may be different, depending on the distribution of the field in the aperture in that plane.

The *directivity* or *directive gain*,

$$G_D = \frac{\text{maximum radiation field intensity}}{\text{average radiation field intensity}} \tag{8.56}$$

The *power gain*,

$$G = \frac{\text{power density in the direction of maximum field intensity}}{\text{power density from an isotropic antenna with the same power input}} \tag{8.57}$$

For a general aperture antenna,

$$G = \frac{4\pi kA}{\lambda^2} \tag{8.60}$$

where k is the efficiency factor and A is the cross-sectional area of the aperture.

The beamwidth approximate relationship

$$\text{beamwidth} \approx \frac{\lambda}{ka} \text{ rad} \tag{8.61}$$

At the end of a waveguide feed, a horn both increases the aperture area and provides an impedance match between the waveguide and the characteristic impedance of the surrounding space. The limiting length for the horn is given by

$$l_{min} = \frac{a^2}{\lambda} + \frac{\lambda}{16} \tag{8.63}$$

Example 8.6 Find an expression for the radiation pattern due to an aperture excited with a cosine distribution of field across the aperture. The distribution is the same as that of the dominant mode in the broad dimension of rectangular waveguide.

Answer The radiation pattern is the Fourier transform of the field distribution in the aperture. The Fourier transform is given in eqn. (8.51) and the field distribution is

$$A(x) = A \cos \frac{\pi x}{a}$$

Let

$$k = (\pi/\lambda) \cos \theta$$

Then substituting into eqn. (8.51)

$$E_\theta = \int_{-1/2a}^{1/2a} A \cos \frac{\pi x}{a} \exp(-jkx) dx$$

Integrating by parts gives

$$E_\theta = \frac{-(A\pi/k^2 a) \cos \frac{1}{2} ka}{1 - (\pi/ka)^2}$$

As the constant term in the expression for E_θ is irrelevant in expressions for the radiation pattern, then

$$\text{radiation pattern} = \frac{\cos[(\pi a/2\lambda) \cos \theta]}{[(a/\lambda) \cos \theta]^2 - 1}$$

Example 8.7 A waveguide horn needs to have an aperture dimension of two wavelengths. Calculate the minimum length of the horn.

Answer The required length is obtained by substituting $a = 2\lambda$ into eqn. (8.63). Therefore

$$l_{min} = \frac{4\lambda^2}{\lambda} + \frac{\lambda}{16} \approx 4\lambda$$

PROBLEMS

8.6 A uniformly illuminated aperture antenna is to operate at a frequency of 3.0 GHz. If the aperture is square, find the dimensions of the aperture to give a gain of 10. What is the beamwidth of the antenna?

[89 mm, 56°]

8.7 As an approximation to the ideal antenna described in Problem 8.6, a pyramidal horn of square aperture, 10 cm by 10 cm, is excited from rectangular waveguide operating in the dominant mode. Find the gain of this antenna. What is the minimum length of the horn?

[10.2, 1.06λ]

8.5 REFLECTOR ANTENNAS

The easiest way in which to increase the area of an aperture antenna is to use a reflector designed on the principles of geometrical optics. The antenna then consists of two basic elements, a relatively small feed and a large reflecting surface. A parabolic reflecting surface has the useful property of transforming a diverging spherical wavefront into a parallel plane wavefront. This is the same principle as that used in the design of optical reflectors such as occur in reflecting telescopes and spotlights or searchlights. The parabolic reflector is a surface of revolution of a parabola characterised by the equation

$$y^2 = 4fx \tag{8.65}$$

which is illustrated in Figure 8.25. f is the distance to the focus. The property of the parabola is that any ray starting from the focus is reflected from the parabolic surface into a ray parallel to the x-axis. What is more, the total distance travelled along each ray path FGH is the same, so that all the points H, H', H", etc., are in phase. For microwave frequencies, the reflecting surface is a metal paraboloid with a small waveguide horn placed on the axis of the paraboloid at the focus. The horn is called the *primary radiator* and the reflector the *secondary radiator*.

In any reflector antenna, the aperture of the reflecting surface controls the radiation pattern and hence the gain, beamwidth and sidelobe levels. The feed horn may be designed to produce any desired amplitude distribution of electromagnetic energy across the aperture of the reflector. The simplest parabolic reflector system gives a uniform illumination across the face of the reflector aperture. In practice many factors modify this ideal result. No horn gives an isotropic radiation pattern so that the radiation amplitude tends to

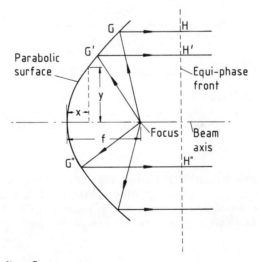

Figure 8.25 Parabolic reflector antenna.

decrease towards the outside of the paraboloid. If the paraboloid has too large a diameter compared to its focal length, the reflecting surface is not properly illuminated by the horn. Conversely, if the paraboloid is too small in diameter compared to its focal length, electromagnetic energy is lost over the edge of the reflector surface as *spill over*.

A support structure is needed to locate the feed horn at the focus. A simple support is a tripod structure but it disturbs the radiation pattern of the reflector by creating a shadow and increasing the sidelobes in the radiation pattern. Any shadow may be eliminated by locating the focus away from the reflected beam, in a system using an offset parabola as shown in Figure 8.26. The reflector consists of only part of the paraboloid not including its axis. Another system which removes the feed antenna from the front of the reflector is the Cassegrain system. Here a hyperbolic subreflector is used to re-reflect the rays from the primary feed onto the parabolic surface of the main reflector, as shown in Figure 8.27. The useful property of the hyperbola is that rays emanating from one focus are focused onto the other. The second focus of the hyperbola is located at the focus of the parabola and provides a virtual point for the primary feed to the parabolic reflector.

The gain of a reflector antenna is still given by eqn. (8.60). Uniform illumination across the aperture of the reflector is the ideal, but for most systems having non-uniform illumination across the aperture, the efficiency factor, k, usually lies between 0.5 and 0.6. A receiving antenna will see a plane wave uniformly illuminating the aperture. However, by the reciprocity principle, if such an antenna had a non-uniformly illuminated aperture when it was used as a transmitter, it will be less than 100% efficient in directing the electromagnetic radiation in its aperture into the receiver. An

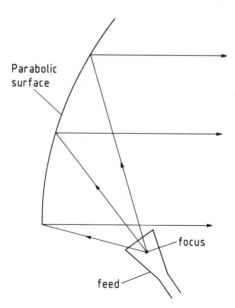

Figure 8.26 Offset paraboloid antenna.

262 RADIATION AND ANTENNAS

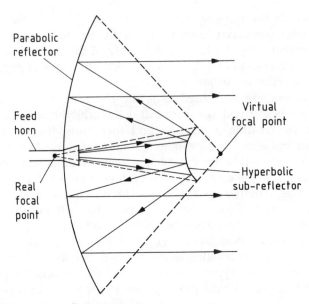

Figure 8.27 Cassegrain reflector system.

effective capture area is defined by

$$A_e = \frac{G\lambda^2}{4\pi} \tag{8.66}$$

Therefore, from eqn. (8.60)

$$A_e = kA \tag{8.67}$$

For lower-frequency operation, the reflector does not have to be a continuous metal surface but a wire mesh surface is satisfactory. Similarly, a waveguide feed and horn would be impractically large and the primary feed is a dipole antenna, or a simple Yagi array. Sometimes different designs are used to control the radiation pattern in two perpendicular planes, for example a simple airfield radar uses an antenna consisting of an offset parabola to control the vertical radiation pattern and an antenna array to control the horizontal radiation pattern.

SUMMARY

A paraboloid is used as a reflector for the electromagnetic radiation and produces a plane wavefront and a parallel beam. The primary radiator, which may be a waveguide horn or a dipole antenna, is placed at the focus of the parabola.

The cross-sectional area of the aperture of the reflector determines the gain, bandwidth and sidelobe levels. The gain

$$G = \frac{4\pi k A}{\lambda^2} \qquad (8.60)$$

Used as a receiver, the reflector antenna has an effective capture area

$$A_e = \frac{G\lambda^2}{4\pi} \qquad (8.66)$$

$$A_e = kA \qquad (8.67)$$

Example 8.8 A satellite communications antenna to operate at 5.0 GHz consists of a parabolic reflector of 1.6 m diameter and has an efficiency factor of 0.6. Calculate the gain of the antenna and its effective capture area when used as a receiving antenna.

Answer The wavelength of operation is given by eqn. (4.31),

$$\lambda = \frac{c}{f} = \frac{3.0 \times 10^8}{5.0 \times 10^9} = 0.060 \text{ m} = 6 \text{ cm}$$

It is assumed that the aperture of the reflecting surface is a circle of radius 0.80 m; therefore the gain is given by eqn. (8.60):

$$G = \frac{4\pi k A}{\lambda^2} = \frac{4\pi \times 0.6 \times \pi \times 0.64}{36 \times 10^{-4}} = 4211$$

Antenna gain is usually quoted in decibels, therefore

$$G = 36.2 \text{ dB}$$

The effective capture area is given by eqn. (8.67):

$$A_e = kA = 0.6 \times \pi \times 0.64 = 1.2 \text{ m}^2$$

PROBLEM

8.8 A satellite television receiving antenna to operate at 12 GHz consists of a paraboloid reflector of 1.0 m diameter. Given that its efficiency factor, k, is 0.65, calculate its effective capture area and its gain.

[0.51 m², 60 dB]

―――― CHAPTER 9 ――――――――――――――――――――――――――

Systems

Aims: The aim of this chapter is to summarise the theory behind the design of those parts of some communications systems which make use of electromagnetic radiation.

9.1 LINE COMMUNICATIONS

Electrical pulse communication on a two-wire line was one of the first commercial applications of electricity. The behaviour of electrical pulses on a transmission line has been discussed in Chapter 2. In the telegraph, the signalling pulses are at a low frequency and propagate satisfactorily on simply supported open-wire lines. One of the first units of resistance was the *mile*, being the resistance of a mile of iron telegraph wire. Where the open-wire line is inappropriate, the telegraph pulses propagate satisfactorily along two-wire cables.

After the invention of the telephone, the low-frequency audio signals were also propagated satisfactorily along open-wire lines. As the telephone became more popular, the circuits increased in number and were sometimes bundled together into multiple pair cables. After the discovery of modulation techniques, carrier-wave telephony was adopted, whereby a number of audio signals were modulated onto different frequency carrier waves on one pair of wires. In order to increase the number of telephone conversations on one communication circuit, the frequency of the carrier wave was increased until the attenuation on two-wire lines became acceptable. Then a cable consisting of a large number of wire pairs, each circuit devoted to a single telephone conversation, could be replaced by a single coaxial cable with all the telephone conversations modulating different carrier waves at different frequencies on one coaxial cable. As the number of communication circuits increased, the carrier-wave frequencies increased and even coaxial cable had too much attenuation and amplifiers were used every so far along the line.

The number of communication circuits continues to increase, so it is then necessary to go to even higher carrier-wave frequencies or to use cables consisting of a number of coaxial lines. Multiple coaxial line cables have been used but an attempt to use circular waveguide for microwave carrier waves has been abandoned because it is uneconomic. However, microwave frequency carrier waves are used for free-space communication links and for satelite communications as described in Sections 9.3 and 9.4. Even higher frequency carrier waves are now available in fibre-optic communication systems as described in Chapter 7. Fibre optic cables are very low loss and the use of amplifiers along a fibre-optic line is unnecessary unless the line exceeds 100 km in length. At various times radio-wave transmission was used to establish permanent communication links in various parts of the world. Some of the principles governing radio wave propagation are described in Section 9.2. Nowadays, undersea optical fibre communication cables are available so that there is no necessity to use radio-wave or microwave communication links except for communication with mobiles or for economic reasons in areas of inhospitable terrain.

There is extreme competition for the use of the electromagnetic spectrum in radio-wave propagation, so that its use for communication purposes will be restricted to applications where there is no alternative, such as communication with mobiles. Most line communications can be provided by optical-fibre cables which have a wide bandwidth and do not use up scarce resources. Existing copper-wire cables will continue to be used for many years but the number of new wire circuits to be installed will be very few. All the principles of transmission line theory as described in Chapter 3 apply to the carrier wave of any frequency propagating along its appropriate line or waveguide. Many of the principles of transmission-line theory also apply to the behaviour of free-space propagation communication links. The rest of this chapter describes applications involving radiation of electromagnetic waves as opposed to their propagation along a line or waveguide.

SUMMARY

Early telegraph communications used low-frequency pulses on open-wire pairs. Telephone uses audio signals on similar open-wire lines or multiple-pair cables.

Many telephone circuits are combined by modulation onto high-frequency communication links such as coaxial cables, microwave radio links or optical-fibre cables.

9.2 RADIO-WAVE PROPAGATION

For communication purposes, all parts of the electromagnetic spectrum, which is illustrated in Figure 4.1, from visible light downwards in frequency are used. In this section, the modes of terrestial propagation of the lower-frequency bands are discussed. The propagation of waves at these frequencies is influenced by the presence of ionised layers in the upper atmosphere. The region in which these ionised layers occur is called the *ionosphere*. They consist of ionised particles from the dissociation of the gases present by the action of ultraviolet and other radiations from the Sun. A number of layers of these ionised gases have been identified and have been labelled with letters of the alphabet as shown in Figure 9.1. The positions and densities of these layers vary with the time of day and the seasons of the year. However, the E and F_2 layers are present most of the time. The theory of electromagnetic propagation through an ionised plasma is complicated and beyond the scope of this book, but the ionised layer has some of the properties of a conductor and reflects electromagnetic wave propagation. However, at frequencies greater than a certain critical frequency, the electrons are unable to follow the rate of oscillation of the wave and normal propagation through the ionosphere is possible. That is why microwave frequencies are used for satellite communication systems and for communication with satellites. The frequency is sufficiently greater than the critical frequency so as to ensure low-loss propagation through the ionosphere under all conditions. The critical frequency is given by

$$f_c = 9\sqrt{N} \tag{9.1}$$

where N is the ionic density, the number of ions per cubic metre.

The propagation properties of an electromagnetic wave in the atmosphere depend on the frequency of the wave. At very low frequencies, $\lambda \sim 20$ km, propagation occurs all round the world in a waveguide formed between the surface of the Earth and the ionsphere. At higher frequencies, the modes of propagation may be separated into three—the *ground wave*, the *sky wave* and the *space wave*. A surface wave, similar to that described in Section 7.1, will

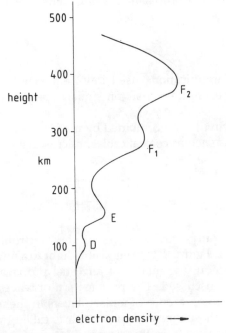

Figure 9.1 Ionised particle density in the ionosphere.

propagate along any surface of a conducting body, provided the wave is vertically polarised, i.e. the electric field vector is perpendicular to the surface. A wave generated by a vertical short dipole antenna close to the surface of the earth is vertically polarised and its magnetic field vector is parallel to the surface. Such a surface wave constitutes the ground wave radiated from any low-frequencies antenna. In theory, a surface wave propagates similarly to a plane wave along a plane surface. If the surface is curved (and any degree of curvature has to be considered relative to the wavelength of the wave) the wave is split, partly following the curvature of the surface and partly radiating in a straight line away from the surface. The finite resistivity of the ground also attenuates the ground wave and limits the range of the ground wave.

There is another effect which sometimes increases the range of the ground wave. That is tropospheric refraction. The troposphere is the region of the atmosphere nearest to the Earth. A direct wave is refracted towards the surface of the Earth because the refractive index of the atmosphere decreases with the decrease in atmospheric pressure which decreases with height. There are other effects due to atmospheric pressure and temperature gradients which can also increase the range of propagation of the direct ray.

Propagation beyond the range of the ground wave is possible by reflection from the ionosphere as shown in Figure 9.2. Such reflected waves are called sky waves. When the sky wave and the ground wave are both received at the same site problems occur due to the different time delay along the two paths leading to distortion of any modulating signal. The critical frequency is also a function of the angle of reflection and vertically directed waves are sometimes not reflected whereas waves at lower angles are. Then there can sometimes be a region with no received signal between the range of the ground wave and the position of the receipt of the nearest reflected sky wave shown in Figure 9.3. This is called the *skip* distance. Multiple-hop sky wave propagation is also possible where a sky wave is re-reflected from the Earth. Again in a region where multiple-hop propagation occurs, the signal may be received simultaneously along a number of different paths leading to distortion. If two signals arrive in antiphase, then they cancel one another

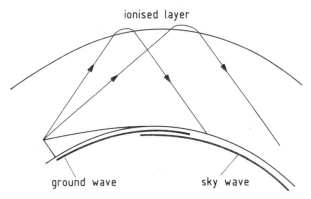

Figure 9.2 Reflection of radio waves from the ionosphere.

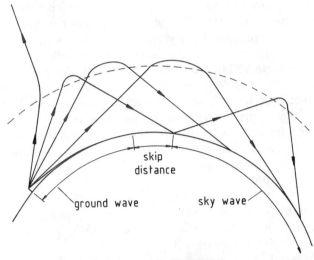

Figure 9.3 Illustrating skip distance and multiple-hop propagation.

out and fading occurs. At higher frequencies, the ground wave is attenuated more strongly and sky wave propagation predominates until the critical frequency is exceeded when point-to-point direct ray propagation only is possible.

SUMMARY

Radio wave propagation is facilitated by the ionised layers in the upper atmosphere in the *ionosphere*.

Very low-frequency propagation occurs around the world in a waveguide formed between the ionosphere and the surface of the Earth.

Medium frequency propagation occurs either by means of a ground wave or by means of a sky wave or both.

The *ground wave* is a combination of the direct ray propagation from the transmitter antenna and a surface wave which follows the curvature of the Earth. The direct ray propagation range is also increased by the effect of tropospheric refraction, whereby the variation of permittivity with height bends the ray towards the surface of the Earth.

The *sky wave* is reflected from the ionosphere and provides useful over-the-horizion communication. The ray may be re-reflected from the surface of the Earth to provide multiple-hop communication paths.

Radiation at frequencies greater than the critical frequency are transmitted through the ionised layer and are not reflected.

9.3 MICROWAVE COMMUNICATION

Frequencies greater than about 30 MHz penetrate the ionosphere and are not reflected back to the Earth. Frequencies appreciably greater than the

MICROWAVE COMMUNICATION 269

critical frequency are transmitted through the ionosphere without much attenuation. Communication using the microwave range of frequencies has to be designed on the basis of direct ray propagation. The high frequencies of the microwave region offer wideband capability allowing simultaneous transmission of many voice and picture channels. They also provide predictable reliable performance which is not always possible with lower-frequency transmissions due to the cyclic and variable nature of the ionosphere. For point-to-point terrestial transmissions, the maximum range is limited by the curvature of the Earth. Consider Figure 9.4. For maximum range, the direct ray path is a tangent to the surface of the Earth. Then by geometry,

$$d_1^2 = (R + h_1)^2 - R^2 = 2Rh_1 + h_1^2 \approx 2Rh_1$$

Therefore

$$d_1 = \sqrt{(2Rh_1)} \qquad (9.2)$$

Tropospheric refraction causes the ray to travel in a curved path towards the surface of the Earth, giving a greater maximum distance, $d_1 + d_2$. This is accounted for by using an increased radius for the Earth. Tropospheric refraction is a variable quantity depending on atmospheric conditions of temperature and pressure. However, for normal conditions of the variation of refractive index with height, the modified radius is given by

$$R' = \tfrac{4}{3}R \qquad (9.3)$$

The radius of the Earth is 6.37 Mm which, when substituted into eqn. (9.3), makes the modified radius to be 8.49 Mm. Then inserting the modified radius into eqn. (9.2) gives an expression for the total range between the microwave transmitter tower and the receiver tower:

$$d_1 + d_2 = (\sqrt{17})(\sqrt{h_1} + \sqrt{h_2}) \times 10^3 \qquad (9.4)$$

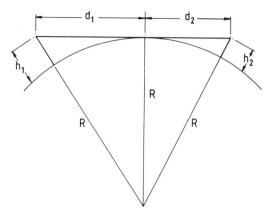

Figure 9.4 Maximum range of a line-of-sight communication link.

The power of ten means that if h_1 and h_2 are measured in metres, the distance is given in kilometres. Microwave communications usually use a line of towers, each at a convenient distance from the last that is less than or equal to the maximum range given by eqn. (9.4). Then each of the intermediate towers has a microwave receiving antenna, an amplifier and a transmitter, which together constitute a repeater. The high value of directivity obtainable from reasonably sized antennas at these frequencies makes such microwave links economic. The effective height of the supporting towers is increased by the height of the local terrain. Consequently, such towers are usually sited on some suitable high ground. For example, a 30 m high tower situated on the ground 200 m above sea level would have an effective height of 230 m provided it were looking across ground effectively at sea level. Then the maximum range between two such towers would be

$$d = 2\sqrt{(17 \times 230)} = 125 \text{ km}$$

The other factor which limits the distance between towers in a microwave communication system is the level of power transmission and reception which can be achieved in a cost-effective installation. The aerials are usually simple front-fed paraboloids with diameters between 2 and 3 m. The gain of the antennas is given by eqn. (8.60). If P_t is the total power radiated, the power density radiated by an isotropic radiator having that total power is given by

$$\text{power density} = \frac{P_t}{4\pi r^2} \qquad (9.5)$$

where r is the radial distance. If the transmitter has an antenna of gain G_t given by eqn. (8.60), the power density at a distance r is given by

$$\text{power density} = \frac{G_t P_t}{4\pi r^2} \qquad (9.6)$$

Equation (9.6) assumes that the peak of the radiation pattern of the transmitting antenna is accurately aligned to the receiver tower. The effective capture area of the receiver antenna is given by eqn. (8.66). Then, if G_r is the gain of the receiver antenna, the power appearing at the terminals of the receiver antenna is given by

$$P_r = \frac{P_t G_t}{4\pi r^2} \times \frac{G_r \lambda^2}{4\pi} \qquad (9.7)$$

If the transmitter and receiver antennas are the same,

$$G_t = G_r = G$$

Equation (9.7) can be rearranged to show the factors influencing the maximum range:

$$r = \frac{G\lambda}{4\pi}\sqrt{\left(\frac{P_t}{P_r}\right)} \qquad (9.8)$$

It is seen that the factors affecting the range, as well as the height and position of the antenna towers, are the power available for transmission and the minimum power detectable by the receiver, the antenna gain and the characteristic wavelength. It is seen that increased antenna gain is more effective than increased power at the transmitter or increased sensitivity at the receiver. However, it is the task of the system designer to decide the most cost-effective solution for any particular installation. For a typical installation, the transmitter power would be 4 W and the received power at the input to the first stage of the receiver amplifier would be 1 μW.

SUMMARY

Maximum range of a microwave link is given by

$$d_1 + d_2 = (\sqrt{17})(\sqrt{h_1} + \sqrt{h_2}) \times 10^3 \qquad (9.4)$$

due to the curvature of the Earth. On an airless planet

$$d_1 + d_2 = \sqrt{(2R)}(\sqrt{h_1} + \sqrt{h_2}) \qquad (9.9)$$

From the power considerations, the maximum range is

$$r = \frac{G\lambda}{4\pi}\sqrt{\left(\frac{P_t}{P_r}\right)} \qquad (9.8)$$

Example 9.1 Each repeater in a microwave communications link has the following characteristics:

Transmitter power, $P_t = 4.0$ W

Minimum detectable receiver power, $P_r = 1.0$ μW

Antenna diameter, $d = 3.0$ m

Antenna efficiency factor, $k = 0.6$

Carrier frequency, $f = 6.0$ GHz

Calculate the maximum spacing possible between repeater towers. Calculate the minimum height of the transmitter and receiver antennas in order to make use of this range capability.

Answer The wavelength is given by eqn. (4.31):

$$\lambda = \frac{c}{f} = \frac{3.0 \times 10^8}{6.0 \times 10^9} = 0.05 \text{ m}$$

The antenna gain is given by eqn. (8.60),

$$G = \frac{4\pi k}{\lambda^2}\left(\frac{\pi d^2}{4}\right) = \frac{\pi^2 \times 0.6 \times 3.0^2}{0.05^2} = 2.13 \times 10^4$$

The range is given by eqn. (9.8),

$$r = \frac{G\lambda}{4\pi}\sqrt{\left(\frac{P_t}{P_r}\right)} = \frac{2.13 \times 10^4 \times 0.05}{4\pi}\sqrt{\left(\frac{4.0}{10^{-6}}\right)} = 1.70 \times 10^5 \text{ m} = 170 \text{ km}$$

The distance between towers is given by eqn. (9.4). Then rearranging eqn. (9.4) gives the minimum height of the antennas:

$$h = \frac{1}{17}\left(\frac{r}{2}\right)^2 = \frac{1}{17}\left(\frac{170}{2}\right)^2 = 425 \text{ m}$$

In this situation, the range of each leg of the communication link is more likely to be limited by the height of the towers than by the characteristics of each repeater.

PROBLEMS

9.1 Taking the transmitter and receiver characteristics and the frequency of operation to be the same as in Example 9.1, calculate the antenna diameter to give a maximum range of 100 km when the antenna efficiency factor is 0.6.

[2.3 m]

9.2 Assuming colonization of the Moon, calculate the maximum range of a microwave communications link whose towers are 36 m in height. Assume that the radius of the Moon is 1.6 Mm.

[23 km]

9.3 The towers of a communications link are separated by a distance of 100 km. The transmitter and receiver antennas mounted on the towers are simple front-fed paraboloids with diameters of 2.0 m and 3.0 m respectively. Each antenna has an efficiency factor of 0.63. Given that the input power to the transmitter antenna is 4.0 W at a frequency of 6.0 GHz, calculate the power at the terminals of the receiver antenna.

[1.43 μW]

9.4 SATELLITE COMMUNICATION

The first suggestion that artificial satellites could be used to provide links between distant points on the Earth's surface was made by Arthur C. Clarke in 1945. He suggested that three satellites, placed in orbit 22 300 miles above the equator and spaced at 120° to one another, could provide complete coverage of the entire planet. Microwave communication links would be established between ground stations and satellites and between satellites. At that height, a satellite orbits the Earth in 24 hours, so it appears to be stationary above one point on the equator. This is just the system we now use with geostationary satellites.

Simple mechanics establishes the conditions governing the orbit of a satellite around any planet. Let the mass of the planet be M and the mass of the satellite be m. The radius of the orbit of the satellite is r. Then from Newton's law of gravitation, the force of attraction is given by

$$f = \frac{GmM}{r^2} \tag{9.10}$$

where G is the gravitational constant. The centripetel force due to the circular motion of the satellite is given by

$$f = mr\omega^2 \tag{9.11}$$

where ω is the angular velocity of the satellite. The two forces given by eqns. (9.10) and (9.11) may be equated to give

$$\omega^2 = \frac{GM}{r^3} \tag{9.12}$$

If the radius of the planet is R and its density is ρ, eqn. (9.12) may be rewritten to give

$$\omega^2 = \frac{4\pi}{3} G\rho \left(\frac{R}{r}\right)^3 \tag{9.13}$$

The gravitational constant, $G = 6.67 \times 10^{-11}$ m³/kg s² and the data for the Earth is $R = 6.37$ Mm and $\rho = 5.5 \times 10^3$ kg/m³. If ω is the speed for rotation once in 24 hours, $\omega = 7.27 \times 10^{-5}$ rad/s, then substitution into eqn. (9.13) gives an expression for the radius of the satellite path:

$$r = 6.26R$$

The orbit of a satellite is usually given as the height above the surface of the Earth. Therefore

$$\text{height} = r - R = 5.26R = 35.8 \text{ Mm}$$

The orbit chosen by Arthur C. Clarke is called the *geostationary* orbit. More correctly it is the synchronous orbit. A satellite in that orbit has an orbital period of 24 hours. A satellite circling the equator in a geostationary orbit in the same direction as the Earth's rotation keeps pace with the Earth and appears to hang motionless in the sky. If the orbital plane is disturbed from the equatorial plane, the satellite describes a figure of eight in the sky with respect to a point on the surface of the Earth. The size of the figure of eight depends on the displacement from the equatorial orbit.

Because of the need to reduce the mass of satellite mounted systems, satellite communication systems tend to use high-power transmitters and large antennas on the ground while limiting the satellite power and the size of the satellite mounted antennas. Satellite communication systems operate in the microwave frequency range. The high frequencies enable narrow beamwidth antennas to be constructed at a reasonable size and ensure loss-free propagation through the ionosphere. Microwave frequencies in particular ensure adequate propagation through cloud and rain, which is not true for infrared and light frequencies. A typical satellite communication link is illustrated in Figure 9.5. For communication with satellites, the ground station antennas need to be narrow beamwidth, but the form of the satellite antennas depends on the application. A particular communication satellite providing a link between two fixed points on the ground requires narrow beamwidth antennas. A satellite communicating with a number of points on the ground requires an antenna providing a much wider coverage on the ground. A satellite used for broadcasting has antennas providing particular coverage on the ground called its *footprint*. As well as providing useful long-distance communication links, satellites are also being used to provide communications in inhospitable and rural areas where the provision of land

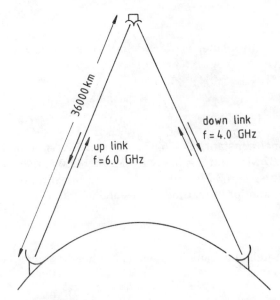

Figure 9.5 Satellite communication link.

lines is uneconomic. Satellites are particularly useful in providing communications with mobiles on land, sea or air.

The power balance equation which was developed for the microwave communications link, eqn. (9.7), also applies to satellite communication systems. However, the range equation given in eqn. (9.8) no longer applies, as it is most unlikely that the gain of the transmitter and receiver antennas are the same. Consider the typical satellite communication link shown in Figure 9.5. Because satellite communications antennas are pointing into the sky, there is very little noise to interfere with the required signal. Therefore the receivers are able to be much more sensitive than those used in terrestrial microwave communication links. A typical receiver is able to operate satisfactorily on a received signal of 10^{-10} W. The range equation is similar to eqn. (9.8):

$$d = \frac{\lambda}{4\pi} \sqrt{\left(\frac{G_t G_r P_t}{P_r}\right)} \qquad (9.14)$$

SUMMARY

The angular velocity of a satellite is given by

$$\omega^2 = \frac{4\pi}{3} G\rho \left(\frac{R}{r}\right)^3 \qquad (9.13)$$

The geostationary satellite orbits at a height of 36 000 km and circles the Earth with a period of 24 hours.

The limiting range for microwave communication is given by

$$d = \frac{\lambda}{4\pi} \sqrt{\left(\frac{G_t G_r P_t}{P_r}\right)} \qquad (9.14)$$

Example 9.2 A satellite in a geostationary orbit has the following on board microwave system: transmitter power is 4.0 W, the antenna diameter is 1.0 m, and the antenna efficiency factor is 0.6. 6.0 GHz is used for the uplink and 4.0 GHz is used for the downlink. The minimum acceptable received signal both at the satellite and on the ground is 10^{-10} W. The system is similar to that shown in Figure 9.5. Given that the same antenna reflector dish on the ground is used for transmission and reception, calculate the minimum antenna diameter required and the transmitter power required for this antenna. The antenna efficiency factor is 0.65.

Answer On the downlink, the frequency is 4.0 GHz. Therefore the wavelength is given by eqn. (4.30):

$$\lambda = \frac{c}{f} = \frac{3.0 \times 10^8}{4.0 \times 10^9} = 0.075 \text{ m}$$

The transmitter antenna gain is given by eqn. (8.60):

$$G_t = \frac{4\pi k}{\lambda^2}\left(\frac{\pi d^2}{4}\right) = \frac{\pi^2 \times 0.6 \times 1.0^2}{0.075^2} = 1050$$

On the downlink, the receiver antenna gain is the only unknown property. Therefore from eqn. (9.7),

$$G_r = \frac{P_r \times 16\pi^2 d^2}{P_t G_t \lambda^2} = \frac{10^{-10} \times 16\pi^2 \times (3.6 \times 10^7)^2}{4.0 \times 1050 \times 0.75^2} = 8.66 \times 10^5$$

The antenna diameter is given from eqn. (8.60):

$$D = \frac{\lambda}{\pi}\sqrt{\left(\frac{G}{k}\right)} = \frac{0.075}{\pi}\sqrt{\left(\frac{8.66 \times 10^5}{0.65}\right)} = 27.6 \text{ m}$$

On the uplink, the frequency is 6.0 GHz. Therefore the wavelength is given by

$$\lambda = \frac{c}{f} = \frac{3.0 \times 10^8}{6.0 \times 10^9} = 0.050 \text{ m}$$

For the same size of antenna, the antenna gain is increased in proportion to the square of the wavelength. Therefore

$$G_t = \frac{8.66 \times 10^5 \times 0.075^2}{0.050^2} = 19.5 \times 10^5$$

The transmitter power is obtained from eqn. (9.7):

$$P = \frac{P_r \times 16\pi^2 d^2}{G_t G_r \lambda^2} = \frac{10^{-10} \times 16\pi^2 \times (3.6 \times 10^7)^2}{19.5 \times 10^5 \times 1050 \times 0.05^2} = 4.0 \text{ W}$$

PROBLEMS

9.4 A satellite in an asynchronous orbit circles the Earth at a height of 1700 km. Calculate its period of rotation around the Earth.

[2.0 hrs]

9.5 A satellite in a geosynchronous orbit at a height of 35 000 km carries a 1.0 W transmitter operating at a frequency of 3.0 GHz, which feeds a 2.0 m diameter antenna. The satellite transmission is received at a ground station which has a 30 m diameter antenna. Assuming that both antennas operate with an efficiency factor of 0.7, calculate the power at the terminals of the receiver antenna.

[88.7 pW]

9.5 RADAR

Another extremely important application of electromagnetic radiation is radar. In particular it is the traditional use of the microwave frequency range. It started in the late 1930s. The name is derived from the initial letters of RAdio Detection And Ranging. It operates by transmitting an electromagnetic wave of sufficient power to ensure that a measurable amount of energy is reflected by the object and detected by a receiver. The simplest form of radar is the pulse radar giving a plan position indication (ppi). It transmits a continuous train of high-power pulses of electromagnetic energy and the range of the object is determined by measuring the time taken for the pulse to travel to the object and return. When operating in air or vacuum, the electromagnetic energy travels at the speed of light so that a $1.0\ \mu s$ round trip delay corresponds to a range of 150 m. The pulse radar usually has a rotating narrow beam like a searchlight and is used for navigation. The direction of the target is obtained from the directional position of the antenna at the receipt of the reflected pulse and its range is obtained from the time delay. This information can be displayed as a radial plot on a cathode ray tube, the ppi.

The CW (carrier wave) or doppler radar gives a velocity indication. It is used in military applications because it is more difficult for an enemy to jam. The doppler radar also has many industrial and consumer uses. As an intruder alarm, it is difficult to eliminate false alarms, such as those from a cat or a curtain moving in a breeze, but it is very suitable for other applications such as controlling a door opener, or operating temporary traffic lights. It is used for the police speed radar and it is being investigated as an anticollision device for vehicles.

Most radars use a common antenna for transmission and reception. The two signals at the input to the antenna feed are separated by a duplexer which prevents the high-power transmission signal from entering the receiver but allows the low-power received signals to enter the receiver unattenuated. The range of the radar is determined by the power of the transmitted electromagnetic wave and the sensitivity of the receiver. Most radar systems operate in the frequency range 1.0–90 GHz. The higher frequency has a shorter wavelength and gives greater discrimination. A good general-purpose radar frequency is about 10 GHz, giving a useful compromise between ease and simplicity of the microwave systems and a reasonable discrimination.

Generally a radar system is required to search for targets and their coordinates. The shape of the radiation pattern depends on the required accuracy of the target coordinates. It will always need to be moved around in two or three dimensions. In most existing systems, this is achieved mechanically but in future scanning will be achieved using electrical means. The beam of a reflector type antenna can be scanned by moving the position of the feed. However, only small angles of scan of about a beamwidth are possible before the radiation pattern deteriorates leading to increased beamwidth and sidelobe levels and loss of gain. Wide-angle electronic scanning is possible where an aperture is filled with a large number of independently controlled sources. Then the radiation pattern beam or beams are controlled by the amplitude and phase of the signal produced at each source.

We shall now calculate the conditions for the effectiveness of a radar system. If P_t is the transmitter power of the radar, the power density of electromagnetic radiation at an object a distance r away is given by eqn. (9.6). If the object is assumed to be a perfectly conducting plane surface, perpendicular to the direction of the radar, of cross-sectional area σ, the electromagnetic power re-radiated by the object is given by

$$P_a = \frac{P_t G_t \sigma}{4\pi r^2} \tag{9.15}$$

In practice σ is defined as the radar cross-section of the object. The power received at the receiver terminals is given by an expression similar to eqn. (9.7):

$$P_r = \frac{P_t G_t \sigma G_r \lambda^2}{(4\pi r^2)^2 4\pi} \tag{9.16}$$

Then the maximum range of the radar is given by

$$r^4 = \frac{P_t G_t G_r \sigma \lambda^2}{P_r (4\pi)^3} \tag{9.17}$$

As for most radar systems the transmitting and receiving antennas are the same antenna, $G_t = G_r = G$, and eqn. (9.17) simplifies to

$$r^4 = \frac{P_t \sigma}{4\pi P_r} \left(\frac{G\lambda}{4\pi}\right)^2 \tag{9.18}$$

Therefore, it is seen that it is much more effective to increase the antenna gain than to increase the transmitted power so as to obtain maximum range. The range is proportional to the square root of the antenna gain whereas it is proportional to the fourth root of the transmitter power.

In the case of a pulse radar, the transmitter power P_t is called the peak power. As the instantaneous power transmitted is unlikely to be constant during a pulse, the peak power is defined as the average of the power during the pulse. However, radar power output is often quoted as an average power with pulses of width τ and a pulse repetition frequency f_r. If the pulse repetition period is T_r,

$$f_r = \frac{1}{T_r} \tag{9.19}$$

and the average power is given by

$$P_{av} = \frac{P_t \tau}{T_r} = P_t \tau f_r \qquad (9.20)$$

In a radar set or any communications receiver, the minimum detectable signal is limited by the noise energy that occupies the same portion of the frequency spectrum as the signal energy. Due to random motion of electrons, every object radiates electromagnetic energy which is called *thermal noise*. The thermal noise also occurs in electrical circuits, where it is directly proportional to the resistance of the circuit. The thermal noise power occurring in any receiver is given by

$$P_n = kTB \qquad (9.21)$$

where k is Boltzmann's constant, $k = 1.380 \times 10^{-23}$ J/K, T is the absolute temperature and B is the receiver bandwidth. At room temperature, $kT = 4 \times 10^{-21}$ W/Hz. However, in any receiver there are other processes which contribute to the total noise in the system, so that the receiver noise is usually greater than the thermal noise. The noise figure of the receiver is a measure of the noise contributed by the receiver circuits. If S_i and N_i are the signal and noise powers at the input to the receiver and S_o and N_o are the signal and noise powers at its output, the receiver noise figure is given by

$$F = \frac{S_i/N_i}{S_o/N_o} \qquad (9.22)$$

In terrestial systems the received signal is competing with noise radiated from the ground or surrounding objects and the minimum detectable signal is quite high. In satellite systems, however, the antenna is directed skywards where all the possible noise sources are at very low temperatures and it is worthwhile to design special low-noise receivers. In such satellite systems it is also useful to have antennas with very low sidelobe levels because sidelobes pick up noise signals from the surrounding ground.

Noise is a random phenomenon and cannot be predicted precisely so that it is necessary for the signal level to be greater than the predicted noise level if it is to be detected with reasonable probability. A radar system or communication system is specified as having a signal level which can be detected with reasonable probability. In a pulse radar system, many pulses are usually returned from any particular object on each radar scan and improve the probability of detection. The imporvement in the signal-to-noise ratio when pulses are integrated is called the integration improvement factor, I_f. A typical integration improvement factor is 10 and typically a 15 dB signal-to-noise ratio is required at the receiver for a 90% probability of detection. Then the minimum signal to noise ratio at the receiver is 31.62. The power P_r in eqn. (9.18) is replaced by

$$P_r = \frac{kTBF}{I_f}\left(\frac{S}{N}\right)_{min} \qquad (9.23)$$

SYSTEMS

SUMMARY

A pulse radar operates by transmitting a pulse and measuring the time delay and direction of any reflected pulse.

The maximum range is

$$r^4 = \frac{P_t \sigma}{4\pi P_r}\left(\frac{G\lambda}{4\pi}\right)^2 \tag{9.18}$$

For a pulse radar, the average power is

$$P_{av} = P_t \tau f_r \tag{9.20}$$

Thermal noise power is

$$P_n = kTB \tag{9.21}$$

The receiver noise figure is

$$F = \frac{S_i/N_i}{S_o/N_o} \tag{9.22}$$

The minimum detectable signal is

$$P_r = \frac{kTBF}{I_f}\left(\frac{S}{N}\right)_{min} \tag{9.23}$$

Example 9.3 A radar has the following specification:

frequency of operation = 3.0 GHz
average power = 250 W
pulse length = 5.0 μs
pulse repetition frequency = 350 Hz
antenna dimensions = 6.0 m by 2.5 m
antenna efficiency factor = 0.7
thermal noise power = kTB = 5.0 × 10^{-15} W
noise factor = 10 dB
integration improvement factor = 10
minimum detectable signal to noise ratio for a 90% probability of detection = 15 dB.
System losses in the feed to the antenna = 3.0 dB.

Calculate the range at which an aircraft with an echoing area of 4.0 m² will first be detected.

Answer First of all it is necessary to change the numbers expressed as dB into ratios. Therefore,

$$F = 10 \text{ dB} = 10$$

$$(S/N)_{min} = 15 \text{ dB} = 31.62$$

$$\text{System losses} = L = 3.0 \text{ dB} = 2.0$$

The system losses further reduce the power available at the receiver terminals, so that eqn. (9.23) becomes

$$P_r = \frac{(kTB)FL}{I_f}\left(\frac{S}{N}\right)_{min} = \frac{5.0 \times 10^{-15} \times 10 \times 2.0 \times 31.62}{10} = 3.16 \times 10^{-13} \text{ W}$$

The wavelength is given by eqn. (4.30);

$$\lambda = \frac{c}{f} = \frac{3.0 \times 10^8}{3.0 \times 10^9} = 0.10 \text{ m}$$

The antenna has a rectangular aperture. Its gain is given by eqn. (8.60):

$$G = \frac{4\pi kA}{\lambda^2} = \frac{4\pi \times 0.7 \times 6.0 \times 2.5}{0.10^2} = 13\,195$$

Then the maximum range is given by eqn. (9.18), using eqn. (9.20):

$$r^4 = \frac{P_{av}\sigma}{f_r 4\pi P_r}\left(\frac{G\lambda}{4\pi}\right)^2$$

$$= \frac{250 \times 4.0}{5.0 \times 10^{-16} \times 350 \times 4\pi \times 3.16 \times 10^{-13}}\left(\frac{13\,195 \times 0.10}{4\pi}\right)^2 = 16 \times 10^{20} \text{ m}^4$$

Therefore

$$r = 2.0 \times 10^5 \text{ m}$$

The aircraft will first be detected by the radar at a range of 200 km.

PROBLEMS

9.6 An air traffic control radar operates at a frequency of 3.0 GHz. It has a common antenna of aperture 6.0 m by 2.0 m, which has an efficiency factor of 0.8. It has a pulse length of 1.0 μs and a pulse repetition frequency of 300 Hz. The transmitter average power output is 150 W. The thermal noise power in the receiver is 4.0×10^{-15} W and it has a noise factor of 10 dB. The loss in the system is 3.5 dB but it also has a pulse integration improvement factor of 9.0. The signal-to-noise ratio required for 90% probability of detection is 14.7 dB. Calculate the range at which there is a 90% probability of detecting an incoming aircraft which has a radar echoing area of 20 m².

[400 km]

9.7 A target tracking radar uses a common antenna of 2.0 m diameter, which has an antenna efficiency factor of 0.7, for both transmission and reception. The pulse power output of the magnetron transmitter is 250 kW at a frequency of 10 GHz. The receiver is capable of tracking an echo in range and angle down to a signal level of 10^{-12} W, but requires a level of 10 dB greater than the minimum level for first detection of a target echo. The waveguides between the magnetron and the antenna introduce a loss of 2.0 dB and those between the antenna and the receiver introduce a loss of 1.5 dB. Calculate the maximum range at which an aircraft of 4.0 m^2 echoing area will be detected.

[66 km]

Bibliography

INTRODUCTORY ELECTROMAGNETIC THEORY

C. Christopoulos, *An Introduction to Applied Electromagnetism*, Wiley (1990).
P. Hammond, *Electromagnetism for Engineers*, 3rd ed, Pergamon (1986).

ELECTROMAGNETIC THEORY

D. K. Cheng, *Field and Wave Electromagnetics*, 2nd ed, Addison-Wesley (1989).
P. Hammond, *Applied Electromagnetism*, Pergamon (1971).
W. H. Hayt, *Engineering Electromagnetics*, 5th ed, McGraw-Hill (1989).
J. D. Kraus, *Electromagnetics*, 4th ed, McGraw-Hill (1991).
S. Y. Liao, *Engineering Applications of Electromagnetic Theory*, West (1988).
C. R. Paul and S. A. Nasar, *Introduction to Electromagnetic Fields*, McGraw-Hill (1982).
S. Ramo, J. R. Whinnery and T. Van Duzer, *Fields and Waves in Communication Electronics*, 2nd ed, Wiley (1984).
K. F. Sander and G. A. L. Read, *Transmission and Propagation of Electromagnetic Waves*, 2nd ed, Cambridge (1986).
J. A. Stratton, *Electromagnetic Theory*, McGraw-Hill (1941).

TRANSMISSION LINES

R. A. Chipman, *Transmission Lines*, McGraw-Hill (1968).
C. W. Davidson, *Transmission Lines for Communications*, 2nd ed, Macmillan (1989).

WAVEGUIDES

A. J. Baden Fuller, *Microwaves*, 3rd ed, Pergamon (1990).
W. S. Cheung and F. H. Levien, *Microwaves Made Simple*, Artech House (1985).
F. E. Gardiol, *Introduction to Microwaves*, Artech House (1984).
P. A. Rizzi, *Microwave Engineering*, Prentice-Hall (1988).
D. Roddy, *Microwave Technology*, Prentice-Hall (1986).
K. F. Sander, *Microwave Components and Systems*, Addison-Wesley (1987).

FIBRE OPTICS

P. Halley, *Fibre Optic Systems*, Wiley (1987).
W. B. Jones, *Introduction to Optical Fibre Communication Systems*, Holt, Reinhart and Winston (1987).

ANTENNAS

F. R. Connor, *Antennas*, Arnold (1972).
M. S. Smith, *Introduction to Antennas*, Macmillan (1988).

APPENDIX 1

Physical Constants

Speed of light, $c = 2.998 \times 10^8$ m/s
Charge on electron, $e = 1.602 \times 10^{-19}$ C
Mass of electron, $m = 9.109 \times 10^{-31}$ kg
Boltzmann's constant, $k = 1.380 \times 10^{-23}$ J/K
Permeability constant, $\mu_0 = 4\pi \times 10^{-7}$ H/m

Permittivity constant, $\varepsilon_0 = \dfrac{1}{c^2 \mu_0} \approx \dfrac{1}{36\pi \times 10^9}$ F/m

Characteristic impedance of free space, $\eta_0 = \sqrt{\left(\dfrac{\mu_0}{\varepsilon_0}\right)} \approx 120\pi = 377\ \Omega$

Radius of the Earth, $R = 6.37$ Mm
Density of the Earth, $\rho = 5.5 \times 10^3$ kg/m^3
Gravitational constant, $G = 6.67 \times 10^{-11}$ m^3/kg s^2

APPENDIX 2

Notation

A	arbitrary constant, network parameter, area
A	vector potential
a	radius, a waveguide dimension, waveguide radius
B	arbitrary constant, network parameter, susceptance, bandwidth
B	magnetic flux density
b	radius, a waveguide dimension
C	capacitance, arbitrary constant, network parameter
c	speed of light, a waveguide dimension
c	as subscript: current, cut-off, critical
D	arbitrary constant, network parameter, a diameter
D	electric flux density
d	diameter, a distance
d	differential operator
E	electric field intensity
e	a distance
e	as subscript: electric
F	noise factor
F	arbitrary vector
f	frequency, a function, focal distance, force
G	conductance, gain, gravitational constant
G	arbitrary vector
g	as subscript: waveguide, group
H	magnetic field intensity
h	height
h	as subscript: magnetic
I	electric current, modified Bessel function of the first kind, integration improvement factor
i	as subscript: input, incident

NOTATION

J	Bessel function of the first kind
J	electric current density
j	$= \sqrt{-1}$
K	modified Bessel function of the second kind
k	wavenumber, an integer, antenna efficiency factor, Boltzmann's constant
L	inductance
l	length, an integer
M	mass of the Earth
m	mass, depth of modulation, an integer
m	as subscript: magnetic, maximum
N	an integer, ionic density, noise power
n	refractive index, an integer
n	as subscript: normal, normalised, noise
o	as subscript: output
P	power, arbitrary constant
p	power density
p	as subscript: phase
Q	electric charge, Q-factor, arbitrary constant
q	line charge density
R	radius, resistance, radius of the Earth
r	radius, range, radial coordinate
r	as subscript: relative, reflected, radially directed component, receiver
S	VSWR, space factor, signal power
S	Poynting vector
s	surface area
s	as subscript: surface, source
T	a time period, pulse repetition period, absolute temperature
$\mathcal{T}$	transmission coefficient
t	time
t	as subscript: tangential, terminating, transmitted, transmitter
U	unit vector
V	electrostatic potential
v	volume, velocity
W	complex dimension
w	energy density
X	reactance
x	linear dimension
x	as subscript: x-directed component

Y	admittance, Bessel function of the second kind
y	linear dimension
y	as subscript: y-directed component
Z	impedance
z	linear dimension
z	as subscript: z-directed component
α	attenuation constant, an angle
β	phase constant
γ	propagation constant
δ	differential coefficient, loss angle
ε	permittivity
η	intrinsic impedance, efficiency
θ	angle, angular dimension
θ	as subscript: θ-directed component
κ	coupling coefficient
λ	wavelength
μ	permeability
π	$= 3.14159$
ρ	electric charge density, reflection coefficient, density
σ	conductivity, surface charge density, dispersion, radar cross-section
τ	time period, propagation delay, pulse length
ϕ	phase angle, power factor, angular dimension
ψ	an angle
ω	angular frequency, angular velocity
∇	vector differential operator
∂	partial differential operator

Index

Absolute potential 6
Absorber, mode 182
Absorption loss 214
A.c. resistance 133
Admittance diagram 92
Aerials 224–263
Ampère 1
Ampère's law 6, 14, 16, 23, 24, 29–31
Antennas 224–263
Aperture antennas 252–259
Arrays of antennas 233–243
Attenuation constant 62, 124, 125
Attenuation of wave 124–127, 214–216
Attenuator
   rotary 182
   vane 171
   variable 222
Automatic network analyser 86

Beamwidth of an antenna 237, 245, 255
Bessel function 174, 175, 177, 207, 208
   zero of 178
Bessel's equation 173
Blooming of lenses 115
Boltzmann's constant 279, 285
Boundary conditions 7, 197
   at waveguide walls see under
       appropriate shape of waveguide
Brewster angle 118
Brewster effect 121
Broadcasting 248, 249
Broadside array 236, 237

Capture area 270
Carrier wave telephony 264
Cassegrain system 261
Cavity see Resonant cavity
Chain network parameters 78

Characteristic impedance see Impedance, characteristic
Characteristic wavelength 109, 135, 137, 139
Charge density 3
Choke coupling 169, 170
Circle diagram 80
Circular fibre 206–213
Circular polarisation 180–182
Circular waveguide 152, 172–186
   boundary conditions 174, 177
   cut-off conditions 174, 177
   dominant mode 180–185
   field components 175–178
Cladded core fibre 207
Coaxial line 147, 148, 150
Communications 264
Conducting medium 128–134
Conduction current 25, 129
Conductivity 4
Connector 216
Conservative field 18
Coulomb's law 29, 30
Coupler, directional 219–223
   polarisation sensitive 222
Coupling coefficient 221
Coupling factor 220
Coupling length 217
Couplings, waveguide 169, 170
Critical angle 199
Critical frequency 266
Curl 14–16
Current 3, 25, 130
Current density 3
Current in waveguide wall 168, 169
Cut-off 138, 150, 158, 174, 177, 189, 203, 209
Cylindrical cavity mode chart 189
Cylindrical polar coordinates 172

Del 11, 14
Dielectric film waveguide 199–205
    fields in 200
Dielectric rod waveguide 155, 206–213
    boundary conditions 208
    cut-off 209
Dielectric slab waveguide 155
Differential operator 11
Dipole aerial 121, 224–232
Directional coupler 219–223
Directive gain of antenna 254
Directivity
    of antenna 255
    of coupler 220
Director 241
Disc resonator 191
Dispersion in optical fibres 216–218
Displacement current 25, 168
Div 12
Divergence 11, 14
Divergence theorem 14
Dominant mode 166–171, 178, 180
Doppler radar 277

Earth, density 273, 285
Earth, radius 273, 285
Effective capture area 262
Effective permittivity 128
Efficiency 74
Efficiency factor 255, 261
Eigenfunction 161
Eigennumber 161
Electric current 3
Electric field intensity 3
Electric flux density 3
Electric probe 189, 190
Electromagnetic field
    components 106–111
    equations 102
    in circular waveguide 175–178
    in conducting media 129, 130
    in free space 106–111
    in rectangular waveguide 163–166
    quantities 2–5
    relationships 5–9
Electrostatic field of antenna 229
Elliptical waveguide 153, 154
Endfire antenna array 238
Energy 25, 26
Equivalent T-circuit 69

Fading 268
Far field of antenna 230
Faraday 1
Faraday's law 6, 14, 17
Ferrite rod aerial 121
Fibre, optical 193–223, 265
Field intensity 3
Field quantities 2–5
Film modes 200
Film parameter 202
Finline 154
Flange 169
Flexible waveguide 169
Flux density 3
Footprint 274
Fourier transform 253
Fraunhofer region of antenna 230
Fresnel zone of antenna 229
Frequency 62
    cut-off 138, 150, 174, 177, 178

Gain of an antenna 255, 261, 270
Gauss's law 5, 6, 12, 14, 29, 31
Gauss's theorem 14, 23, 26
Generalised frequency parameter 202, 209
Geostationary orbit 274
Gilbert 1
Glass 126, 133
    fibre 211
Grad 11
Graded index fibre 211
Gradient 10
Gravitation 273
Gravitational constant 273, 285
Greenhouse 126
Grounded antennas 248, 249
Ground wave 266, 267
Group velocity 143, 144, 216
Guided film modes 200

Half silvered mirror 133
Helical wave 181, 182
Helix 180–182
Hertz 1
Horizontal polarisation 121, 250
Horn 252, 256
Hyperbolic reflector 261

INDEX

Image waveguide 155
Impedance
   characteristic 35, 39, 60, 64, 70, 114, 148, 169
   diagram 80
   effective 79, 80, 86
   input 80
   intrinsic 108, 114, 125, 230
   matching 91–99
   measurement 86–90
   mismatch 256
   normalised 79
   of free space 108, 285
   open-circuit 69
   short-circuit 69
   source 48
   transformation 92
   transformer 92
   transmission line 148
Index of refraction 118, 199, 200
Induction field of an antenna 229
Inductive loading 65
Infrared absorption 215
Interaction length 221
Intrinsic impedance 108, 114, 125, 230
Inverse square law 29
Ionic density 266
Ionosphere 265–267
Irrotational conservative field 18

Junction 48

Kirchhoff's current law 36
Kirchhoff's voltage law 8, 36

Laminations 132
Laplacian 13, 172
Laplace's equation 13
Left-hand circular polarisation 182
Light, speed of 4, 34, 64, 104, 285
Line communications 264, 265
Line efficiency 74
Line termination 41–45
Linear antenna array 233
Linear polarisation 121, 172, 180, 181
Long wire antenna 244–251
Lorentz condition 225
Loss, attenuation 124–126
Loss tangent 125, 126

Magnetic field intensity 3
Magnetic flux density 3
Magnetic probe 189, 190
Main beam of an antenna 236
Marconi 1
Matched junction 92
Matched termination 92
Matching 91
Matching section 171
Material dispersion 216
Material properties 3, 4, 285
Maxwell 1, 25
Maxwell's equations 23–28, 29, 101, 106, 148, 156, 163, 175, 195, 198
   solution of, in a conducting medium 128–130
   solution of, in a non-conducting medium 101–104
Microstrip 149, 150, 191
Microwave communications 268–272
Mode 140
   absorber 182
   chart 190, 210
   conversion 217
   dispersion 216
   dominant 166–171, 178, 180
   film 200
   guided film 200
   HE_{11} in dielectric rod waveguide 209
   nomenclature 140, 161, 189
   radiation 200
   space 200
   substrate 200
   TE 140
   TE in circular waveguide 178
   TE_{11} in circular waveguide 178–185
   TE in parallel plate waveguide 140, 141
   TE in rectangular waveguide 164
   TE_{10} in rectangular waveguide 166–171
   TE in thin film waveguide 200, 202
   TE surface wave 197
   TEM 140
   TEM in coaxial line 148
   TEM in parallel plate waveguide 140
   TM 140
   TM in circular waveguide 175
   TM_{01} in circular waveguide 184
   TM in parallel plate waveguide 140, 141
   TM in rectangular waveguide 164, 165
   TM in thin film waveguide 201, 202

Mode (*cont.*)
  TM surface wave 197
  transformer 182
  transmission line 136, 140
  waveguide 136, 140
  waveguide in coaxial line 150
Modified Bessel function 209

Nabla 11
Near field of antenna 229
Negative circular polarisation 182
Neper 62
Network analyser 86
Newton's law of gravitation 273
Noise 279
Noise figure 279
Normalised frequency 209
Normalised impedance 79
Notation 286
Numerical aperture 200, 207

Oersted 1
Offset parabola 261
Ohm's law 4
Open-circuit impedance 69
Optical fibre 193–223

Parabolic index optical fibre 211
Parabolic reflector 260, 261
Parallel plate waveguide 135–141
Parasitic antenna 240, 241
Permeability 3
  complex 124
  constant 3, 104, 285
  relative 4
Permittivity 3
  complex 124
  constant 3, 104, 285
  effective 128
  relative 3, 200
Phase
  constant 63, 104, 124, 125
  velocity 62, 64, 104, 142, 143, 167, 216
Physical constants 285
Plan position indicator 277
Plane coupling 170
Plane of polarisation 121, 182
Plane wave 101–134, 230
  field components 106–111
  in conducting medium 128–134
  in non-conducting medium 101

Poisson's equation 13, 31
Polarisation 121, 140, 172, 180–182, 250
Polarisation sensitive coupler 222
Positive circular polarisation 182
Potential 6, 18, 19, 30, 224–226
Power 25, 26, 231
  gain 255
  loss 131, 132
Poynting vector 26, 230
Primary radiator 260
Propagation constant 60, 63, 104, 124, 125, 128
Pyramidal horn 256, 257

Q-factor 186, 191
Quarter-wave matching section 92

Radar 238, 240, 277–281
Radar cross-section 278
Radiation 224–263
  field of antenna 229
  modes 200
  pattern 230, 236, 238, 239, 241, 245, 248, 253
  resistance 231
Radio receiver 121
Radio wave propagation 265–268
Range 269, 271, 275
Rayleigh scattering 215
Rectangular waveguide 152, 153, 156–171, 254
  boundary conditions 159, 165
  cut-off conditions 158, 160, 165
  dominant mode 166–171
  field components 163–166
  flexible 169
  wall currents 168, 169
Reciprocity principle 241
Reflection coefficient 43, 48, 70, 71, 79, 80, 86, 91, 114, 119, 120
Reflection from a plane boundary 112–123
Reflector 241
Reflector antennas 260–263
Refraction 112–123
  index of 118, 199, 200
Repeater 270
Resistive film attenuator 171
Resistivity 4, 191
Resonance curve 187
Resonant antennas 245

Resonant cavity 186–192
Resonant frequency 187
Resonator *see* Resonant cavity
Retarded potentials 225, 226
Rhombic antenna 247, 248
Ridge waveguide 153
Right-hand circular polarisation 182
Rotary attenuator 182, 183
Rotating joint 184

Satellite 109, 273–276
Scalar potential 6, 18, 30
Scattering loss 214
Screw matching section 171
Secondary radiator 260
Sectoral horn 256, 257
Semiconducting glass 133
Separation of variables technique 157, 173
Sheath loss 215
Short dipole antenna 224–232
Short-circuit impedance 69
Shorted stub 91
Shunt stub 92
Sidelobes 236, 237
Single mode optical fibre 210, 217
Skin depth 132, 133
Skip distance 267
Sky wave 266, 267
Sliding load 87
Slotline 150
Slotted line 170
Smith chart 80, 81, 86
Snell's law 117
Solenoidal field 18
Source impedance *see* Impedance, source
Source reflection coefficient 48
Space factor 236, 239
Space modes 200
Space–time diagram 49–56
Space wave 266
Special theory of relativity 34
Speed of light 4, 34, 64, 104, 285
Spherical polar coordinates 227, 228
Splice 216
Spill over 261
Standing waves 69–76
Step index fibre 213
Stokes's theorem 16
Stripline 148, 149
Stub 91
Stub matching 91

Substrate modes 200
Surface current 130, 133
   density 8
Surface mode 154
   boundary conditions 197
Surface wave 193–198
Switching surges 34–58

T-circuit equivalent of transmission line 69
Taper 182
Telegraph 264
Telephone 264
Telephony 264
Television 249
Termination *see* Matched termination
Thermal noise 279
Thin film waveguide 199–205
   electromagnetic fields in 200
Time domain reflectometry 88
Top loading an antenna 249, 250
Total internal reflection 120, 198, 199
Transformer core 132
Transmission coefficient 43, 114, 119
Transmission lines 146–151
Transmission line 34–100, 136
   a.c. effects 58–100
   constants 36, 38, 39
   equations 36–41
   equivalent T-circuit 69
   fields 146–151
   impedance 39
   lossless 37
   mode 136
   two-conductor 146, 147
Transverse attenuation parameter 202
Transverse phase constant 202
Travelling wave antenna 246
Triplate line 149
Tropospheric refraction 267, 269
Two-wire transmission line 147

Uniform field 17
Unit vector 11

Vane attenuator 171
Vector
   analysis 10–23
   differential operator 10, 11
   operator identities 17
   potential 19, 30, 224, 226
   voltmeter 86

Velocity
  group 143, 144, 216
  of light 4, 34, 64, 104, 285
  phase 62, 64, 104, 142, 143, 167, 216
  wave 38, 61, 64, 142–145
Vertical polarisation 121, 250
Voltage step 45–57
Vortex 18
VSWR 71, 86, 114, 170

Wall currents 168, 169
Wave, plane *see* Plane wave
Waveguide 136, 152, 252
  circular *see* Circular waveguide
  dielectric rod 155, 206–213
  dielectric slab 155, 199–205
  dispersion 216
  elliptical 153, 154
  mode 136, 140, 150
  parallel plate 135–141
  rectangular *see* Rectangular waveguide
  ridge 153
  thin film 199–205
  wavelength 138, 140
Wavelength 61, 63
  characteristic 109, 135, 137
  cut-off 138, 139, 150
  waveguide 138, 140
Wave propagation 101–105
Wave velocity 38, 61, 64
Weakly guiding films 201

Yagi array 241, 242